CRYOGENICS

Helen Hale, Editorial Consultant
Illustrations by Gruger Studio, Inc.

CRYOGENICS

by RICHARD J. ALLEN

J. B. LIPPINCOTT COMPANY
PHILADELPHIA AND NEW YORK

Contents

1 CRYOGENICS 7

2 USES OF CRYOGENIC LIQUIDS 22

3 CRYOBIOLOGY AND CRYOMEDICINE 37

4 REFRIGERATION 48

5 THERMOMETRY, INSULATION, AND CRYOSTATS 73

6 CHANGES IN PROPERTIES OF MATTER AT LOW
     TEMPERATURES 92

7 LIQUID HELIUM AND SUPERFLUIDITY 103

8 SUPERCONDUCTIVITY 112

9 CRYOGENIC ELECTRONICS 126

10 PHYSICAL RESEARCH 147

   INDEX 159

Chapter 1

CRYOGENICS

THE word *kryos* was used by the ancient Greeks to mean *icy cold*.

To these people living in a land noted for mild sunny winters, *kryos* could scarcely have meant very low temperatures. It might have been used to describe their summer drinks which the rich cooled with carefully stored ice or snow. It might also have been used to describe early Greek attempts at food preservation. But however *kryos* was used in ancient times, it was a far cry from its use in 1880 when scientists working on ways of obtaining lower and lower temperatures used it in forming the word *cryogenics*. They used Cryogenics to refer to the field of extremely low temperatures, a field of science which is now becoming very important to us.

But just what is this low temperature region? What are its uses? How do we cool things to these low temperatures and what happens to them when we do? What strange effects are found which do not exist at room temperature? And how does Cryogenics affect us and our way of life now and in the future? These are some of the questions which will be answered.

Although Cryogenics is nearly a hundred years old, it is only in the past few years that it has grown from a field in which scientists were working in the laboratory to learn more about the laws of nature at low temperature to include many areas outside the laboratory. While some industrial applications of Cryogenics were found many years ago, most of their work was pure scientific research. Practical applications are now growing very rapidly.

In the world that we know and see around us, where are these low temperatures used? One very dramatic example is found in our current space experiments, for most large missiles and space vehicles depend on cryogenic products to propel them. Crews in bombers vital to our strategic air defense breathe oxygen that earlier was a cryogenic fluid; and the atomic bombs carried by these giant planes were first converted into the deadlier hydrogen bombs through the addition of cryogenic liquids.

Industry relies on cryogenics to furnish certain gas requirements; steel mills, for example, use hundreds of tons a day of liquid oxygen (obtained from the air through the use of cryogenics). New electronic devices such as masers and lasers (Microwave Amplification by Stimulated Emission of Radiation and Light Amplification by Stimulated Emission of Radiation) are cooled by cryogenics. Cryogenics is used in agriculture, and in medicine during surgery and to preserve blood and biological specimens. It is surprising to realize not only how rapidly and widely the application of this field of science has grown, but also how much every one of us depends on cryogenics without even being aware of it.

When we think of the cold, or of cold temperature, we usually have in mind a feeling of discomfort, whether it describes a temperature that is lower than that of our bodies, or an atmosphere too cold to live in. Icy Cold suggests temperatures well below freezing, somewhere near zero degrees

Fahrenheit, or the coldest day we can remember, or perhaps a chilled drink so cold that the outside of the glass is frosted. But even the coldest day ever recorded on the earth—125°F below zero in Antarctica on August 25, 1958, did not descend even to the *hot* end of the cryogenic temperature range.

There is no clearcut ceiling to the cryogenic region, though many people define it to be the point where air becomes so cold that it turns into a liquid. This takes place at about 330°F below zero, so there is not much doubt that the scientists who first defined cryogenics as the *field of low temperature* meant this to be the field of *very* low temperature. However, there are some interesting low temperature effects which happen above 330°F below zero, but below those which we can reach with our ordinary refrigerators and freezers, and so today we sometimes go beyond this definition and include this higher range of temperature.

While the warm end of the cryogenic region may not be clearly defined, there is no question of what we mean by the low end of this temperature range. Cryogenics extends down to the lowest temperature which is predicted by science, a temperature of almost 460°F below zero (−459.63° to be precise), which is known as *absolute zero.*

We are all directly concerned with temperature through the effect it has on our comfort. Just a small change from what we consider ideal requires radical changes in our dress and our activities. In fact, it does not take much of a change to make life impossible. Yet, despite this great dependency which we have on temperature, do we really know what temperature is? Most of us are like the people Mark Twain was describing when he said, "Everybody talks about the weather, but nobody does anything about it." We use the term temperature frequently, but temperature is something we never actually feel or measure directly. Instead we actually feel or measure properties of matter which change with temperature.

There are two ways in which temperature can be described. The way that we normally think of temperature is *relative* temperature. However, if we try to describe temperature in terms of just what is producing it, and how it is related to our other units of measure, such as length, weight, time, and energy, then we are seeking the *physical* definition.

Relative temperature is a thermal state which considers whether the substance will absorb heat from another substance, or give up heat to it. An object is hot if it gives up heat to a second object; cold if it absorbs heat from this second object. We measure temperature in this relative sense by placing a thermometer in contact with whatever we want to measure, such as your tongue when you are ill. If you have a fever, your tongue will be hotter than the thermometer and will transfer your heat to the thermometer. Similarly, a thermometer placed outdoors when it is cold will go down, since it yields up its heat to the cold air. When heat transfer is completed, the thermometer is at the same temperature as the substance we are trying to measure, whether it is your body heat, the winter cold, or the water in the ocean. What we actually measure is the change that has occurred in the thermometer.

It is more difficult to understand what is meant by *physical* temperature. To do so we must examine the structure of the material whose temperature we wish to measure, and the atoms and molecules that make up this structure. It is necessary to think of temperature in terms of the average internal energy of the molecules. This concept may not be very clear, but if we look a little further the connection be-

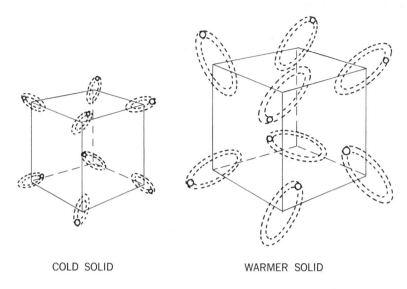

COLD SOLID　　　　　　WARMER SOLID

Figure I-1. The motion of atoms increases as temperature increases.

tween temperature and the structure of matter will become evident.

Even at low temperatures, the atoms and molecules which make up all matter are never still. It is true that in a solid, atoms and molecules are not able to move around freely; they are arranged in regular patterns that make up their crystal structure (Figure I-1). But while they cannot leave their place in this pattern, they are still free to move back and forth within their assigned positions. Newton's Second Law of Motion, one of the most basic laws of physics, and the relation to kinetic energy (which is based on that law) tell us that the kinetic energy of a moving object is dependent upon its mass and speed. Since the atoms and molecules in matter are moving, they must have kinetic energy because of this motion (kinetic means moving). The average amount of this internal kinetic energy of motion on the part of many molecules

11

determines *physical* temperature. Reduced to its simplest terms, the more the temperature is raised, the greater the movement of molecules. The relationship between temperature and average internal molecular energy is only accurately applicable to a gas, and then only under certain conditions. But because the difference between it and a solid or a liquid is not too great, we can make use of it to understand and describe temperature, even though it cannot be used to measure temperature.

From what has been described so far, we can see that the temperature increases when the average molecular kinetic energy increases. By this we mean that additional energy must be added to the substance, such as the heat it receives when touching a warmer body, in order to cause its temperature to rise. And from this develops a very important physical concept, namely that *heat* is a form of energy.

When we add heat, that is energy, to a cold substance, the temperature and kinetic energy of the atoms and molecules increase. This means that the speed at which they move must increase, causing the atoms to travel further away from their center positions. They then bump into each other and interfere with each other's progress. Some of this added energy helps move their center positions in the crystal lattice further apart. The result is that the material or crystal made up of tremendous numbers of these atoms and molecules whose spacing is increasing will also increase in size. If we understand this, we can see clearly why it is that metal, for example, expands with increased temperature. This is why expansion joints must be used in building bridges and concrete highways.

Let us study these atoms and molecules even further (Figure I-2). We remember that when they are cold, their movement

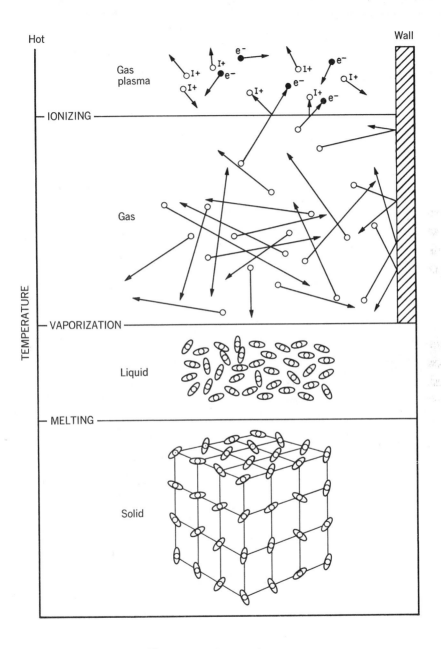

Figure I-2. States of matter.

is small. As their kinetic energy, temperature, and speed increase, they eventually reach a temperature where force between molecules is no longer strong enough to hold them in their permanent center positions or in any orderly crystal lattice. They then become free to wander about through the substance. These wandering molecules form a liquid, and the temperature at which this takes place is the melting point. As their temperature and speed continue to increase, the molecules finally gain enough energy to overcome the remaining forces still holding them together in the liquid, and they escape completely from the matter. Thus is called the vaporization point, because the molecules, on escaping (evaporating) have formed a gas.

Something else occurs now that these molecules are essentially independent of each other and are moving in all directions, colliding with the walls of the container in which they are held. Pressure results from the force generated by the many million collisions that take place each second on the container walls. If the temperature increases, the energy and motion of the molecules does also; these collisions by faster and faster moving particles become more and more frequent and so the pressure rises.

If we add more and more energy, or heat, to these atoms now in a gaseous state, we find that eventually they move so rapidly that one or more of the electrons in the outermost orbit of the structure will become separated from the rest of the atom. This atom is then said to be *ionized*. The electrons and ionized atoms grouped together are known as a plasma. When they recombine to form un-ionized atoms, the extra energy is given off in the form of light. This is how the neon sign lights work, for example.

As scientists worked on the nature of matter, on energy,

pressure, and other problems having to do with temperature, they realized that they needed some standard method of measuring temperature, and so they established and defined scales. One of the first of these men was a German instrument-maker named Gabriel Daniel Fahrenheit. In 1724 he made a glass thermometer in which he used mercury as the temperature sensing fluid. Mixing together salt, ice, and water, Fahrenheit established his zero point. Mixing ice and water alone, he obtained a second point, and called it 30°, since this was an easily divisible number. Body temperature came at 96° on his scale and the boiling point of water at 212°. Later on, Fahrenheit discarded the salt, ice, and water measurement, and to make the scale a little more simple, he redefined his freezing point of water as 32°. This gave the instrument a total scale of 180° between his melting and freezing points of water. This is the thermometer most commonly used in the United States and Great Britain.

In much of the rest of Europe, however, and for most scientific work requiring temperature measurement, the scale used is called Centigrade. Although the origin is not certain, credit is usually given to a Swedish scientist, Andreas Celsius in 1742. He placed the boiling point of water at zero, and the freezing point at 100°, the exact reversal of what we know as the Centigrade scale. But since both are based on the properties of water, and the two reference points are always at the same temperature, at atmospheric pressure, these scales can always be accurately reproduced. On the Centigrade scale, sometimes called the Celsius scale, anything colder than the freezing point of water is measured in minus degrees.

There are many other scales used to measure temperature, and indeed we could certainly do as Fahrenheit and Celsius did and make one up ourselves. We could pick two events which occur in nature, such as the melting or boiling points of some substance. We would mark some number to note the colder of these two points, probably at zero, and another number to mark the warmer point, say at the convenient figure of one hundred. When we divide the space between these two points into even divisions, we have marked out the degrees on our scale.

It must be noted that scientists working on the problems of measuring temperature must be careful in the choice of points of reference for their scales. It must be possible to repeat these points at any time, and they must always remain the same. The attempt by Fahrenheit to use body temperature as one of these reference points demonstrates this need for caution, since our temperature can not only change from hour to hour, but from person to person.

All students learn at some time or other how to work out the conversion of temperature from Fahrenheit to Centigrade or the reverse. It is quite simple, and involves only the use of this formula:

$$\text{T FAHR.} = \left[\text{T CENT.} \times \frac{180° \text{ F}}{100° \text{ C}} \right] + 32° \text{ F} = \frac{9 \text{ T CENT.}}{5} + 32° \text{ F}$$

There are two other scales used quite often in the measurement of temperature, known as the Rankine and Kelvin Scales. These are similar to Fahrenheit and Centigrade in terms of the size of the degrees used, but they are both absolute scales. This means that the zero degree marked on them is at absolute zero—the lowest temperature science can predict (Figure I-3).

The term Absolute Zero has been used several times and has been referred to as the lowest temperature which is pre-

16

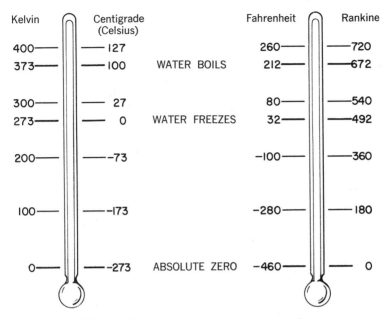

Figure I-3. Common temperature scales.

dicted by theory. We cannot say that it is the lowest temperature which could ever be reached, since the theory also says that there is no way of attaining absolute zero even though scientists have reached 0.00001 degrees Kelvin.

We can describe this as the temperature at which atoms, molecules and their electrons are at the lowest energy they can possibly possess. Let us try an experiment, to see if it will make this definition more meaningful.

Imagine that we take a bottle which contains a quantity of gas, such as helium or argon. This bottle is equipped with a pressure gauge, and a way of changing the temperature by precise amounts.

Suppose we start our experiment with the temperature of

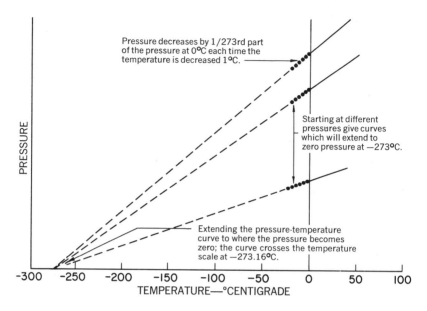

In the figure:

Pressure decreases by 1/273rd part of the pressure at 0°C each time the temperature is decreased 1°C.

Starting at different pressures give curves which will extend to zero pressure at −273°C.

Extending the pressure-temperature curve to where the pressure becomes zero; the curve crosses the temperature scale at −273.16°C.

PRESSURE

TEMPERATURE—°CENTIGRADE

Figure I-4. A method for determining Absolute Zero.

the bottle and its gas at zero degrees centigrade. If we can decrease the temperature by one degree, we find that the pressure reading decreases by $\frac{1}{273}$ of its value at $0°C$. If we decrease the temperature to two degrees below $0°C$, the pressure decreases by $\frac{2}{273}$ of its $0°$ value, and so on. From this we conclude that a constant amount of gas at a temperature near $0°C$ undergoes a decrease of $\frac{1}{273}$ of its pressure at $0°$ centigrade for each degree decrease in temperature. This experiment can be shown also on a graph (Figure I-4).

Let us extend the decreasing pressure down; it reaches zero pressure at $-273°C$. Since the pressure can decrease no fur-

18

ther we conclude that the temperature cannot decrease any more either, and therefore that −273°C must be Absolute Zero. If we repeat our experiment with a different gas, or start at another pressure on our graph, the lines still all cross the axis at −273°C.

In describing this experiment, we have ignored certain things which we would have to consider if we were actually carrying our gas down to very low temperatures. For one thing, the gas would become a liquid before we reached absolute zero. Secondly, at temperatures where our gas had not yet become liquid, it might not always have a pressure decrease of $\frac{1}{273}$ of the 0°C pressure each time the temperature is lowered one degree. We had to assume that we had a perfect or ideal gas that followed the laws governing gases. But scientists know how to correct actual gases so that difficulties such as these are taken into account. If this is done, our experiment can be used in determining absolute zero quite accurately. If we examine these gas laws we find the scientific explanation for the method of determining absolute zero.

In conducting our experiment we used the Centigrade Scale. Let us, however, now express temperature in the absolute units of the Kelvin Scale. If we assume a pressure value, we can use the experiment as an example of the other gas laws.

Suppose that at 273°K (0°C) the pressure gauge reads 546 units. If we divide this pressure (P_1) by the temperature (T_1) we get $\frac{P_1}{T_1} = \frac{546}{273} = 2$. We use the number after P and T to show that this was our first measurement. After the temperature is lowered one degree to 272°K, the pressure decreases by $\frac{1}{273}$ of 546, which is 2 units. The pressure reading for this second measurement (P_2) is 544 units. Dividing this pressure

19

(P_2) by the temperature (T_2) again gives us the answer of 2:

$$\frac{P_2}{T_2} = \frac{544}{272} = 2.$$

From this we can see that however many measurements we make at lower and lower temperatures, the pressure divided by the temperature for a constant volume remains constant:

$$\frac{P_1}{T_1} = \frac{P_2}{T_2} = \text{Constant.}$$

The law which we just described through equations and which we followed in determining absolute zero was first arrived at by Joseph Louis Gay-Lussac in 1802. This law bears the name of this French chemist and physicist who also discovered a second law that relates the volume of a gas to its temperature while at a constant pressure:

$$\frac{V_1}{T_1} = \frac{V_2}{T_2} = \text{Constant.}$$

The first of the gas laws was discovered earlier, in 1660, by the English scientist Robert Boyle. Boyle's Law relates pressure and volume of gas at constant temperature:

$$P_1 V_1 = P_2 V_2 = \text{Constant.}$$

This shows that the pressure multiplied by the volume of a certain amount of gas always gives the same value if the temperature does not change. By combining the laws of Boyle and Gay-Lussac, we arrive at the General Gas Law:

$$\frac{P_1 V_1}{T_1} = \frac{P_2 V_2}{T_2}$$

If any of the three quantities (temperature, pressure, and

20

volume) does not change between the two measurements, it may be omitted from both sides of the equation and we are left with one of the previous laws.

Perhaps the most important thing for us to learn and remember about these laws is that they use absolute values of temperature. Whenever temperature is used in a physical law, it is expressed in units of an absolute scale. Because non-absolute scales can use as zero any point which the definer wishes to set, and because this zero point usually has no connection with the physical law, we can see that in scientific terms the temperature arrived at is meaningless. This is why the absolute scales were defined and why a knowledge of what absolute zero means is important even in introductory physics and chemistry courses.

We have now gone over many of the basic ideas involved in understanding the region of Cryogenics: the state of matter, definitions of temperature, and absolute zero. But this region of low temperature is not of interest only in itself. Since man is normally quite practical, he is usually interested in something because he feels that he can use it to his advantage. There are many practical applications of the science of cryogenics. After describing some of them, we will turn to the means by which low temperatures can be produced and maintained, to the unusual properties of materials at these low temperatures, to such interesting physical phenomena as the superfluidity of liquid helium, and superconductivity.

Chapter 2

USES OF CRYOGENIC LIQUIDS

WE KNOW that gas is a state of matter as are the liquid state and the solid state. When the temperature of a substance which we normally think of as being a gas is lowered, the substance will become a liquid or if it is lowered even further it may enter the solid state. This holds true for all gases, except helium, and so includes oxygen, hydrogen, neon and even the air we breathe. (Helium will become a liquid, but will not become a solid at atmospheric pressure no matter how low the temperature goes.) A list of the temperatures at which common gases liquefy (which is also their boiling temperature) is given in Table II-1.

The substances which are liquid in the cryogenic temperature range are often called cryogenic fluids. Since we are most familiar with these fluids in the gaseous state at room temperature, we will often use the self-contradicting term "liquid gas." A large cryogenic industry has grown up around these fluids and their cryostats, which is the name given to the insulated containers in which they are stored.

One of the best-known areas in which cryogenic fluids are used, and certainly one of the most dramatic, is the field of rocketry. All of us have seen photographs of large rockets at

TABLE II-1

BOILING TEMPERATURES OF COMMON SUBSTANCES

If the substance is being cooled, this is the liquefication temperature. If it is being warmed, it is the vaporization temperature. In either direction the change in state occurs at the same temperature which for simplicity is called The Boiling Point Temperature.

SUBSTANCE	BOILING TEMPERATURE AT ATMOSPHERIC PRESSURE		
Water	100°C	373.2°K	212°F
Ethyl Alcohol	78	351.2	154
Methyl Alcohol	65	338.2	149
Ether	35	308.2	95
Chlorine	− 35	238.1	− 31
Xenon	−107	166.1	−161
Krypton	−153	120.1	−244
Methane	−161	111.7	−259
Oxygen	−183	90.1	−297
Argon	−186	87.3	−302
Fluorine	−188	85.0	−307
Nitrogen	−196	77.4	−320
Neon	−246	27.1	−411
Deuterium	−250	23.6	−417
Hydrogen	−253	20.3	−423
Helium −4	−269	4.2	−452
Helium −3	−270	3.2	−454

the moment of launching and doubtless we noticed the frost on the outside of the rocket section that housed the liquid oxygen. Many of us watched on television as one of our astronauts was propelled into space. During those final tense seconds the first movement we were aware of was the cloud of white vapor outside the rocket. This is visible when the liquid oxygen is boiling off. The rocket cannot depend upon the atmosphere to provide the oxygen needed to burn the fuel it carries, so it must carry its own oxygen supply. To carry this huge amount of oxygen in a gaseous state would either require carrying a large low-pressure container that would be bulky and difficult to lift, or thick-walled, high-pressure metal containers which would also be very heavy. But in liquid oxygen there are about 700 times as many molecules in the same space then there would be if it were a gas at the same pressure. This means that the liquid oxygen can be carried efficiently at atmospheric pressure in lightweight insulated containers. Because of this advantage, liquid oxygen, or LOX as it is nicknamed, has been used in rockets since the first practical long-range rocket was built. This was the V-2 Rocket, used to such effect by Germany near the end of World War II.

This rocket was first made much larger than a man, stood 46 feet high, and when loaded with fuel weighed 28,500 pounds. It was developed by German scientists (including Wernher von Braun) near the town of Peenumunde on the Baltic Sea. It carried 8,000 pounds of alcohol for fuel and 11,000 pounds of liquid oxygen to support the combustion. This gave the rocket a range of about 200 miles. Hitler boasted of his new secret weapon and believed it would be the means to ultimate German victory. Some 4,300 V-2's were fired by the Germans in the final stages of the war. About 1,200 landed within the city limits of London and killed or injured 8,000 people. Londoners found the rockets far more terrifying than the worst of the bombing raids.

The construction of the V-2 is representative of many of the rockets which were to follow. The liquid oxygen and the fuel were piped from their separate storage tanks and through the valves and controls to the rocket engine combustion cham-

German V-2 rocket being prepared for a night launching. This captured rocket was being tested at White Sands, New Mexico, in 1946. (U. S. Army photo)

ber. Since then there have been many different design improvements. Some rockets have used the fuel and LOX chambers to form the skin of the rockets to save weight, for example. Others used the LOX pipeline to cool the walls of the combustion chamber.

In spite of the various improvements, almost all long-range rockets since the V-2 have carried liquid oxygen. Recently some solid and room temperature liquid rocket propellants have been developed which provide their own oxidizer, such as that used in the Polaris missile. But while this appears to decrease the need for cryogenic fluids, research in propulsion for very large rockets has led to the use of liquid hydrogen as the propellant. Both liquid hydrogen and liquid oxygen will be carried in this rocket. When they are combined and burned they form water. The tremendous energy released in this reaction will be used to propel the rocket.

Plans for nuclear-propelled rockets also include carrying liquid hydrogen and oxygen. Here again, they would be combined to form water as in the liquid hydrogen propelled rocket, but in addition, the reactor would then heat the water vapor to a very high temperature. The water vapor would leave the rocket with a much higher velocity than that achieved by the burning of the hydrogen and oxygen alone and give a far greater thrust for the same amount of fuel.

A nuclear reactor cannot propel a rocket by itself despite the tremendous energy it can release. This becomes evident if we examine the means by which a rocket moves. A rocket moves forward by pushing material backward, in accordance with Isaac Newton's Third Law of Motion. This law states that for every force there is an equal and opposite force. When we apply this law to the rocket, we see that the force due to the mass of the rocket, multiplied by the change in velocity, is

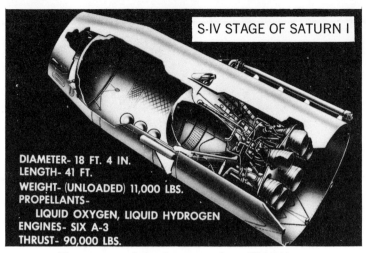

Two stages of the Saturn rocket. (NASA)

equal and opposite the force due to the mass of the fuel which has been burned and ejected in the opposite direction, multiplied by its change in velocity. To obtain maximum thrust with the least possible mass of fuel, the rocket designer searches for ways by which the mass may be ejected at a very high velocity, such as by the rapid burning and ejection of fuel in ordinary rockets. Liquid hydrogen and oxygen give a large

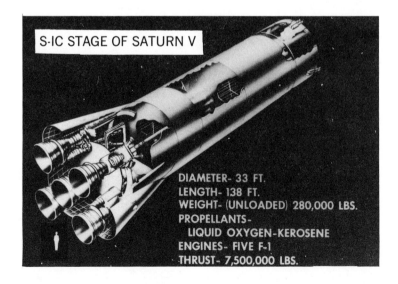

amount of thrust when burned. By heating them further with the reactor, the velocity of the ejected gas is increased and the thrust is increased accordingly. A nuclear rocket power plant is being developed and tested in Nevada. An idea of the rate at which the liquid hydrogen fuel is used can be obtained from the size of the cryostat in which the liquid hydrogen is stored which holds almost a quarter of a million gallons.

Just as the use of cryogenics provides a convenient way of carrying gases in rockets, it also proves to be a convenient way of transporting gas for many other needs. For example, many large cryostats have been built on trucks and even railroad cars for transporting industrial gases in liquid form. The extent of the improvement resulting from shipping the gas as a liquid can perhaps be seen from the fact that with the high-pressure metal cylinders, it takes about 100 pounds of steel to hold a single pound of gas, while with the cryogenic liquid, it is often a pound or less. Transportable cryostats make it possible to ship all the cryogenic fluids by truck, from liquid oxygen and nitrogen down to liquid hydrogen and helium.

Turning back for a moment to the history of rocketry, those German scientists who developed the V-2 rocket also had to transport the liquid oxygen to their missile sites. To do this, they built their liquid oxygen plants underground where they would be safe from bombing and carried the LOX to the launching sites by train and trucks. Railroad cars were built which each carried 48 tons of LOX. This was then transferred to trucks carrying five to eight tons to complete delivery.

Since those days much more design and development has gone into rail and truck transportation of cryogenic fluids. Today it is possible to transport over 28,000 gallons of liquid

Liquid hydrogen storage cryostat at Mercury, Nevada. (Kaiser Aluminum Co.)

hydrogen from Florida to California in one of the largest railroad tank cars ever built. In spite of the shaking and bumping that occurs during the nine-day trip, less than 3% of the hydrogen has evaporated when it arrives at its destination.

Helium, which is quite expensive, has also been found to be practical to ship long distances by truck or even by airplane. A new source of helium was recently discovered in Saskatchewan, Canada. Since this source is so far from the places where the helium would be used, it will be necessary for the helium to be liquefied (at 4.2°K) and trucked or flown out to its markets, if it is to be economically practical to extract and sell it. Already many of the companies which pro-

Liquid hydrogen can be transported successfully in one of the largest railroad cars ever built. This car can transport 28,000 gallons. (Linde)

duce liquid helium for laboratories ship it by air to their customers.

The transportation of liquefied natural gas by ship has now been proved practical and will soon provide a new source of fuel for Europe. Natural gas consists mostly of methane, which boils at 112°K; only slightly above the temperatures used to define the cryogenic region. The gas will be transported in ships which are actually ocean-going cryostats.

Natural gas is obtained from oil wells. In the United States this gas is saved and used since pipelines are available to carry it to the consumer. But in other parts of the world, such as the Middle East and North Africa, oil fields are far removed from the consumer, and the gas is usually burned where it is found and thus wasted. This happens even though countries such as England and France, with their limited sources of fuel, must rely on industrial gas obtained from coal for their requirements.

The transportation of liquid methane from Texas to England as an experiment proved so successful that facilities are now being built in North Africa to trap and liquefy natural gas from the oil fields there. Two large methane tankers, the *Methane Prince* and the *Methane Princess,* have been built with a capacity of about 12,000 tons of methane each. Twenty-nine round trips a year between England and Port Arzew, Algeria, have been projected for these ships. The gas they transport will cost 30 to 50% less than the gas produced in Britain from coal. The French expect to begin importing liquid methane shortly, and plans are even being made to supply Japan with gas from Alaska.

The construction of these cryostat tankers is very interesting. These ships each have nine aluminum tanks for the methane. The tanks rest on balsa wood timbers which not only provide support but also serve as insulation. In addition, the sides of the tanks are insulated with fiberglass and balsa wood. The total insulation results in an evaporation rate of less than 1% per day of the liquid which can be carried. Even this boil-off gas is not wasted, but is burned in the ships' boilers for fuel.

We have described the importance of cryogenics in the field of missiles and our space program, and also its value in the transportation of gases. But we have not yet considered how these gases were obtained to start with. Many were extracted from mixtures of several gases found in nature, by processes involving the use of cryogenics. The most important of these processes is undoubtedly separation of oxygen and nitrogen by means of the liquefication of air. Air separation by liquefication has been the chief means of producing oxygen since before 1900.

If we look back at our Table of Boiling (or Liquefication) Temperatures, we see that oxygen becomes liquid at 90.1°K while nitrogen remains a gas until cooled to 77.4°K. So if the air is cooled to temperatures in this region, the oxygen will become liquid first and can be separated from the nitrogen.

The steel industry is one of the major users of oxygen. It is used for many purposes; for improving the efficiency of furnaces and for purifying the molten metal. In recent years it has been found practical to build air-liquefication and air-separation plants adjacent to the steel mills in order to satisfy their tremendous requirements, rather than to transport the oxygen either as a gas or liquid by truck. It is not uncommon for air-separation plants to produce several hundred tons of oxygen a day. This liquefied oxygen is stored in huge cryostats, but when needed is distributed throughout the steel mill in pipelines as a gas.

In addition to producing oxygen, these liquefying plants

also produce large amounts of liquid nitrogen. Nitrogen too has value as an industrial gas. One of its major uses, again in the steel industry, is in providing a chemically inert atmosphere for heat treating steel.

But liquid nitrogen is also very useful in providing low temperature environments in which to study and preserve a wide range of items, to insulate lower temperature cryogenic equipment, and to machine or process materials as would be impossible at room temperatures. Trucks and trains carrying fresh produce and meats long distances to market are now using evaporating liquid nitrogen to provide refrigeration.

Argon is the most common of the miscellaneous gases which make up 1% of the air; its boiling or liquefaction temperature lies between those of oxygen and nitrogen. It is usually liquefied with the nitrogen where it is allowed to remain as an impurity. If the nitrogen must be pure, or if the argon is desired for other uses, it can also be separated. Krypton and neon as well as oxygen, nitrogen and argon are obtained from the air by liquefaction.

Storage of gases as cryogenic liquids has been found practical in many cases and cryostats in the 1,000- to 5,000-gallon size are now in use. These cryostats are installed permanently on the grounds of a factory and are kept filled with deliveries by cryostat trucks. Hospitals store liquid oxygen in cryostats for use in oxygen tents. Breathing oxygen for military aircraft is often carried as a liquid. Pressure regulators and electrical heaters are used to insure that the pilot and crew have an adequate supply of gaseous oxygen as it is needed. Work is even being conducted to develop skin diving apparatus which would use cryogenic air storage to give increased capacity and to avoid using the large, heavy, high-pressure tanks which are now required.

One of the methods of obtaining a very high vacuum is through the use of cryogenic fluids. This is especially true of the very large vacuum chambers built to simulate both the high vacuum and low temperature conditions of outer space, and also in much laboratory and industrial vacuum equipment. The method by which cryogenics is used to produce a high vacuum is called cryopumping or cold trapping. Cryopumping is a recent term and usually refers to the use of liquid helium. Cold trapping is most frequently applied to the use of liquid nitrogen.

The principle is the same for both. Let us see how it works and how it can be applied. When a surface is made very cold, gas molecules coming into contact with it will condense and remain on the surface. We can understand this if we think of the water vapor in the air. At night it condenses on the cool ground and forms dew. While the ground only cools enough to condense water vapor, a colder surface does much more and will condense gases which have lower boiling temperatures. This is where the cryogenic fluids come in. They are used to cool surfaces in a vacuum system at very low pressures. The cryogenic fluid container is built into the vacuum system so that the vacuum of the system surrounds the container walls. If liquid nitrogen is used to fill the chamber it will naturally cool its walls and reduce to $77°K$ the temperature of the surface in contact with the vacuum region. When the atoms which still remain in the vacuum space come in contact with the $77°K$ surface, those gases with higher boiling temperatures such as argon, oxygen, water vapor, for example (Table II-1), condense and remain on the surface. With these atoms effectively removed from the vacuum chamber the pressure decreases further. The liquid helium cryopumping system does the same thing but is able to produce even lower

34

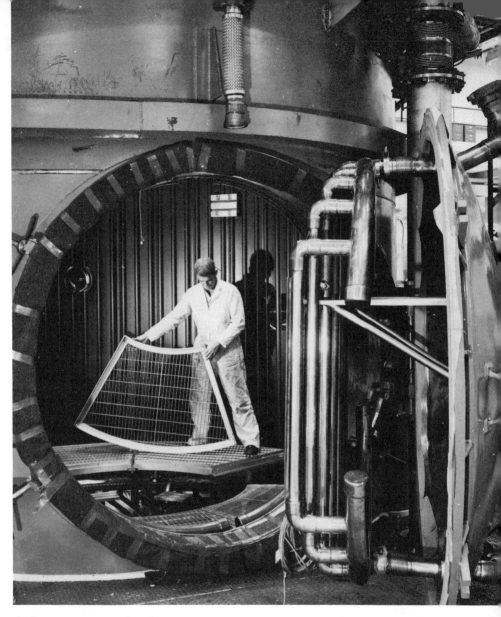

A large vacuum chamber used by NASA for simulating the high vacuum and low temperature of outer space. Liquid nitrogen pipes can be seen inside the door and lining the walls. (Martin Co.)

vacuums. The very low temperature of liquid helium makes it a very good cold trapping substance since all other substances become liquid or solid at various points on the temperature scale above 4.2°K. By condensing all substances but helium, lower pressures can be reached than with liquid nitrogen trapping. However, the higher cost and difficulty in handling liquid helium make its use impractical for most applications.

These are some of the uses of cryogenic fluids. We will examine others after we have looked at some of the low temperature properties of materials and other physical phenomena which occur at low temperature.

Chapter 3

CRYOBIOLOGY AND CRYOMEDICINE

LET US turn from the area of the physical sciences and take a look at how cryogenics is employed both to preserve food and living substance, and in the field of medicine generally. Its use in surgery ranges from the removal of warts to delicate brain operations performed on conscious patients.

Long before man developed machinery to produce low temperatures, he saw that the cold could be used to preserve his food. Farms had springhouses where meat and dairy products were kept cool by evaporating water. Ice was also used for preserving food during the warm summer months. And now, with today's modern household refrigeration, not only is the spoilage retarded, but through the use of low temperature freezers, food can be preserved almost indefinitely.

Just what is it about cold temperatures that preserves food? Quite simply, the activity and rate of growth of bacteria in the food slows up. At room temperature, some of these bacteria act to break down the chemical structure of the food. This alters it both physically and chemically, usually detrimentally, and results in spoilage. If we take food out of the refrigerator and allow it to stand, the temporarily slowed bacteriological processes will start up again and the food will eventually spoil.

From this we see that refrigeration did not kill all the bacteria; some apparently were not damaged permanently but were only slowed down by the cold. This poses an exciting question, for if this can be done to bacteria by cooling, what, then, will occur if we cool other living substances? Can they be put in a state where time stands still, then brought back to normalcy sometime later by allowing them to warm up? If this can be done for bacteria, can we do it with cells, with portions of living creatures, small reptiles, animals, and ultimately with man himself?

The field of cryobiology, that is, the science of freezing in relation to living organisms, takes us to the upper end of the cryogenic region. We cannot use temperatures as high as the freezing point of water, because it has been found that many biological reactions can take place below $0°C$. In fact, some species of micro-organisms have been actually found to multiply at temperatures almost as low as $-9°C$. An example of this biochemical activity below $0°C$ is the change in flavor of some foods after long storage in a freezer, which indicates that some bacteria remained alive and active. Another reason why we maintain the temperature considerably below the freezing point in cryobiology is that ice crystals can damage a biological specimen by destroying its cell walls. As this ice crystal growth has been observed almost as low as $-130°C$, it is desirable to maintain the storage temperature even below this point to insure that such specimens do not deteriorate in storage. As this is only $66°$ (C or K) above the temperature of liquid nitrogen ($77.4°K$ or $-196°C$), which we called the warm end of cryogenics, we seem to be approaching the cryogenic region quite rapidly. It is for practical rather than biological reasons that we preserve the biological specimen at cryogenic temperatures. First of all, there do not seem to be any good refrigerating substances between −130°C and the temperature of liquid nitrogen. Nitrogen itself has certain properties which make it highly desirable as a refrigerant, and so it is the substance commonly used for this purpose.

Liquid nitrogen, secondly, is a relatively inexpensive refrigerant, and it is easy to obtain. It is chemically inert, and it will not burn, explode, or leave a residue when it evaporates; it is, therefore, easy to handle. Its chemical inertness also prevents it from reacting with any specimen stored within it, whose acidity or alkalinity it does not affect.

It is the actual rate of cooling that is most important in the preservation of food. The main object is to stop the effect of bacterial action by suspending the growth of bacteria or by destroying them, if possible without breaking down cell walls. Very rapid cooling to well below the freezing point is the most effective method. This gives us our quick frozen foods, which

A biological storage cryostat for preservation of specimen and cultures at 77° K. (Linde)

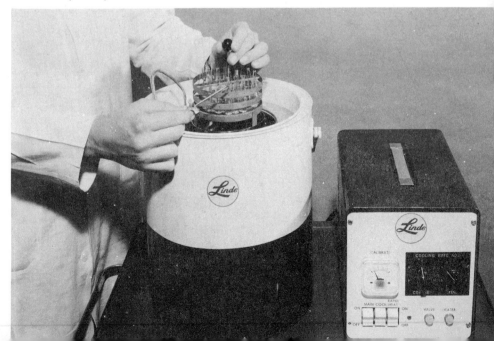

By injecting a fine stream of liquid culture into liquid nitrogen, it freezes almost instantly into separate granules. This preserves the culture and permits any quantity to be removed as needed.

as we know remain in good condition for long periods of refrigeration.

Let us see how a liquid specimen can be preserved in the laboratory. We will fill a hypodermic with our specimen, a culture used for medical purposes. The fluid in the syringe is then sprayed or injected into a flask containing liquid nitrogen; the nitrogen, you remember, boils at approximately 77°K or −196°C, and so it is very cold indeed. As it hits the nitrogen, the culture freezes almost instantly into separate lumps or granules. In this way the specimen can be preserved and granules measured out and used as needed.

Before we turn to the more prosaic aspects of quick freezing food to preserve it, let us detour by way of Siberia. The Beresovka Mammoth, perfectly preserved in ice, was discovered there around the turn of the century. The mammoth was in appearance much like an Indian elephant, but with very long upward-curving tusks and a long thick hairy coat. The scientists investigating this discovery found the animal's flesh was well preserved and actually still fit to eat. Fresh grass and flowers were in its stomach and a buttercup lay unchewed on its lip. Since buttercups have never been native to arctic regions, it is now believed that the mammoth must have been grazing in a warm climate when he was so suddenly and rapidly frozen.

What strange events took place that can explain this prehistoric animal's extraordinary preservation? The theory scientists offer is this: that a heavy cold mass of air developed above the low-lying, warm atmosphere in which the mammoth stood, like Ferdinand, peacefully eating flowers. At first the warm atmosphere remained trapped under this heavy cold layer. But a weak spot developed and through it the cold air suddenly came pouring down. Tremendous winds blew up and dropped the temperature to well below the freezing point in a few minutes. The mammoth must have been killed and frozen immediately. This is not so strange when we realize that even today in Siberia it is quite common for the temperature to drop 60 degrees in a few hours. Our unfortunate mammoth was literally "frozen" in its tracks, and was then covered with snow and ice, to remain quick frozen for 10,000 years— the period of time determined through the radioactive carbon dating process.

There are a number of difficulties, however, in the freezing processes of cryogenics when they are applied both to food and to biological specimens. Suppose we have a specimen of a tissue, which we wish to preserve by freezing. The cells in this specimen contain water in which various chemicals are dissolved. As our cooling process begins, the water starts to freeze. That which remains as a liquid in each cell becomes less though it nevertheless contains all the dissolved chemicals. As the solution becomes stronger or more concentrated, damage to the specimen may occur. Eventually some of the dissolved chemicals may solidify and separate from the solution, thus changing the chemical character of what remains. In addition, the formation of ice crystals, which we mentioned earlier, may do mechanical damage to the cell walls.

Scientists are learning how to solve these problems by careful control of the cooling rate and by the addition of certain protective chemicals to the specimen before cooling begins. Glycerol and glucose (sugar) both reduce the harmful effects of cooling. This use of glycerol was not an original idea of the cryogenic biologist, for there are certain insects who do the same thing when cold weather is coming on. These insects produce glycerol in their bodies, and so are able to change their body fluid and adapt it to the cold. It is along the same lines as changing the oil in a car, and adding antifreeze to the radiator as winter approaches.

There are at least 70 types of tissues and cell types which have been successfully preserved with cryogenics and the protective additives. The more important of these are blood,

Container holding frozen whole blood being removed from storage at liquid nitrogen temperature. (Linde)

bone marrow, semen, various cultures and viruses and pathological samples. Cryogenics has made an important contribution to the preservation of whole blood. Before cryogenic techniques were developed only blood plasma, in many cases less desirable than whole blood, could be stored for long periods. Whole blood could be stored for about three weeks. But whole blood, properly cooled to the temperature of liquid nitrogen, has been kept for over two years without deterioration. What was necessary for the successful preservation of this whole blood? The development of special techniques and cooling equipment so that the blood could be cooled to 77°K without damage. A recent method of preserving whole blood, developed by Dr. Charles Huggens at the Massachusetts General Hospital in Boston, also uses low temperatures and is expected to be more economical. Huggens' process which involves adding chemicals to the blood, does not require temperatures as low as those produced by liquid nitrogen. But though the blood could be successfully preserved at this temperature, the plastic sacks in which it is stored become brittle.

The preservation of bone marrow by cryogenic techniques may develop along similar lines. Ways have been found to replace bone marrow damaged by irradiation whether by excessive X rays or by nuclear accidents. At present it is only possible to use a person's own marrow, and this must be stored in advance, but as research progresses it is expected it will be possible to distinguish marrow types like blood types, and to establish marrow banks.

By freezing cultures and tissues, the biological and medical investigators can keep a wider variety of specimens on hand without worry about their deterioration with time. This is also important in preserving strains of bacteria or yeast without change, as these often undergo changes as they grow and reproduce.

Although living tissues and some larger biological specimens have been preserved by means of cryogenics and recovered without damage, we are still a long way from being able to preserve man. If this were possible, it would be very important in our space travel program, since it will take a long time (perhaps several years) to reach some of the further planets. If our space traveler were properly frozen, he would not need food or water. He could do without entertainment or space to exercise in. And his rate of aging would be greatly slowed up throughout his trip. While the idea is intriguing and might perhaps seem very desirable, it is at present fantasy only, or science fiction. It will undoubtedly remain this way for many years, if not forever.

It is true that French scientists cooled the heart of a baby chicken to liquid nitrogen temperature and successfully made it beat again when it warmed up. Experiments have also been carried out with hamsters. But in spite of this progress, there is still a long way to go and many, many problems to overcome before the first astronaut volunteer can be quick frozen for his trip into space.

Although cryogenic medicine and surgery are relatively new, here too, rapid progress has been made. Dry ice has been in use for some time and now liquid nitrogen has found its place both in the doctor's office and in the operating room.

Liquid nitrogen is very satisfactory in the treatment of the skin, and is used to reduce scars, remove warts, and even to clear up severe acne. The treatment of scars and warts is quite simple. The area to be treated is touched with a cotton swab which has been dipped in liquid nitrogen. The doctor watches to see that this spot turns white, indicating it is frozen, and then removes the swab. This takes only from about ten seconds to a minute. In about six hours a blister forms. The treatment may have to be repeated once a week for several weeks for large warts. When the blister heals after a couple of weeks the defect is usually gone and the scar remaining is very small. It is, apparently, much less painful than other means of wart and blemish removal and is successful in about 90% of the cases. Its use in scar tissue and blemish removal has been found to be better than the "sandpaper surgery" done at higher temperatures.

Cryogenics made its appearance in surgery in a very remarkable operation. The patient was suffering from Parkinson's

Disease, in which there was uncontrollable movement of the arms, legs and other parts of the body. We know this is caused by defects in the portion of the brain which controls the movement. What was always so difficult was the determining of the exact spot on the brain responsible for the trouble, so it could be destroyed without also destroying healthy parts.

A successful method for doing this was one using cryogenics developed by Dr. Irving Cooper of St. Barnabas Hospital in New York, in 1962. Dr. Cooper first made a hole in the patient's skull, thus exposing the general area of the brain responsible for the defect. He then moved over the brain surface a slender wand or probe through which flowed cool nitrogen gas. The probe used by Dr. Cooper consisted of three tubes, one inside the other. It works like this: the cold nitrogen flows toward the probe tip through the innermost tube. Here it cools a piece of silver which forms the probe tip and which is used to cool the brain. After cooling the tip, the nitrogen returns back up the probe between the innermost and middle tubes. The space between the middle and outer tubes is evacuated and serves to insulate the probe so that only the silver tip is cold. Despite the complexity of this instrument, it is at its largest, only 1/10 of an inch in diameter. With the cold silver tip, Dr. Cooper caused the brain cells it was cooling to cease functioning temporarily. His patient, however, remained conscious and assisted the doctor by reporting his reactions. With the cells responsible for the defect located in this manner, the surgeon then lowered the temperature of the gas coming through the probe. This lowered the temperature of the silver tip and froze the defective cells, destroying them permanently, while leaving the remainder of the brain unaffected.

Cryogenics is also being used in agriculture for many purposes, chief of which is the preservation of semen for artificial insemination of cattle and horses. Another interesting use is in the preservation of pollen. One example of this application is the attempt to cross pollinate the slow-growing, good wood-producing northern pine tree with the fast-growing, but poorer wood-producing southern pine. Since these trees germinate in different seasons, and their pollen is short lived, they could not be cross pollinated to produce a better strain. Pollen is now being stored in liquid nitrogen to save it until the cross pollination can be completed.

The probe used to numb portions of the patient's brain. This causes the brain to cease functioning temporarily and thus permits locating defective cells. (Linde)

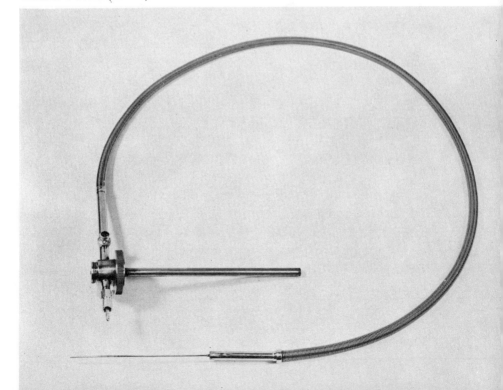

Chapter 4

REFRIGERATION

In ORDER to use cryogenics, man must be able to produce low temperatures when and where he needs them. So we can see that the history of cryogenics follows very closely the history of refrigeration. Each new development in refrigeration made available lower temperatures than were possible before, or made existing low temperatures easier to obtain and maintain and at less expense.

In ancient times, low temperatures were provided only by nature. The early Egyptians had no word in their language for ice; in their climate such a thing did not exist. But they did develop a porous kind of pottery which they filled with liquid, covered with dampened leaves, and put in a windy spot. The liquid from the leaves and that which seeped through the pottery, evaporated, and cooled the remaining liquid to about 5°C below room temperature (in cooler countries ice was even made in this way). Low temperatures for cooling purposes were pretty much of a luxury, however.

Even in less torrid Greece and Rome, only the rich could afford ice or snow in summer. In colder countries, use was made of ice and snow that was gathered in winter and stored underground or in pits until summer. The use of a "cold house" for preserving food was recorded around 2000 B.C. Ice served as the world's principal means of refrigeration until very recently; it was harvested on lakes and ponds in the winter and stored in large wooden "ice houses" until used. In 1805, Francis Tudor, who lived in New England, even established a profitable business of transporting natural ice by ship to southern cities and to islands in the Carribean.

The first machine which actually made ice was built over two hundred years ago. Although scientists soon reached lower and lower temperatures with laboratory techniques (English physicist Michael Faraday, for instance, reached —110°C in 1845), the machinery for producing low temperature for commercial usage developed quite slowly. In 1862, over a hundred years after the first ice machine was built, an improved machine was designed which burned one pound of coal for every four pounds of ice it produced.

In 1851 an American physician, John Gorrie, invented the first mechanical refrigerator patented in the United States. Gorrie lived and practiced medicine in the hot climate of Florida. He was trying to find a way of cooling sickrooms artificially, so that the effects of fever would be reduced. But he lacked money to develop his machine, which was similar to later successful ones. Others were luckier, and in 1868 the first artificial ice plant was built in New Orleans.

Refrigeration development progressed rapidly in the 1870's and 1880's. It was Jonas Linde who in 1876 developed the ammonia refrigerator and the carbon dioxide machine. Ice-producing machines had reached such a stage in development that they were installed in two ships in Glasgow, Scotland, for transporting meat. Scientists reached 48°K in 1884 by using solid nitrogen. By 1898 James Dewar had liquefied hydrogen and reached 15°K. In 1900, the only gas which

49

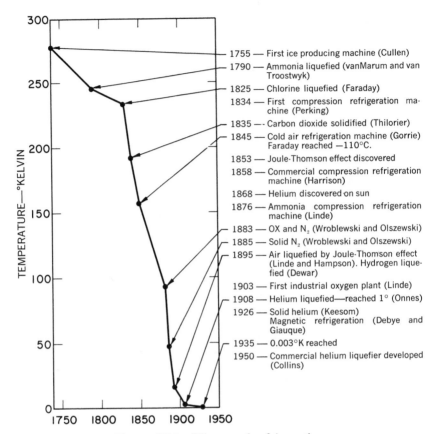

Figure IV-1. History of refrigeration.

had not yet been cooled into liquid form was helium. Not only does helium liquefy at a very low temperature indeed but not much of this gas had been found. Helium had been discovered in the atmosphere of the sun back in 1868, but none was found on the earth until 1895, and then only in very small amounts. It was not until 1905 that helium in large amounts was discovered in Kansas mixed with natural gas. In 1908, a Dutch physicist named H. Kammerlingh Onnes succeeded in lique-fying helium at 4.2°K, and then on the same day achieved a temperature of 0.7°K. In 1935, a temperature of 0.003°K was reached. In 1950, Samuel Collins of the Massachusetts Insti-tute of Technology, designed a commercial helium liquefier which made liquid helium (boiling point 4.2°K) available on a large scale.

The past few years have seen extensive development of cryo-genic refrigeration devices of all kinds, from small systems designed to cool an area no bigger than a dime for use in missile guidance systems, to large systems producing hundreds of tons of liquid oxygen and nitrogen daily.

We can make use of many physical effects to produce low temperatures, and devices have been developed to put each of them to use. Refrigerators have been designed using many different combinations of both these various effects and their various devices. Obviously, we cannot study and describe them all. Fortunately, to understand cryogenic refrigeration, this is not necessary. We will find one basic device which illustrates the application of each of the widely used effects, because most devices can be combined to show the basic refrigeration sys-tems. The large remainder of more refined devices are usually tailored to specific refrigeration uses or are designed to pro-vide more efficient operation, and add little to our basic under-standing of cryogenic refrigeration.

We can boil water at sea level, at 212°F (100°C); if we take it up in the mountains it boils at a lower temperature. In Denver, which is a mile above sea level, water boils at 201°F (94°C). The boiling temperature change that we see take place is due to the decrease in atmospheric pressure as we go to a higher altitude; this pressure is 14.7 pounds per square inch

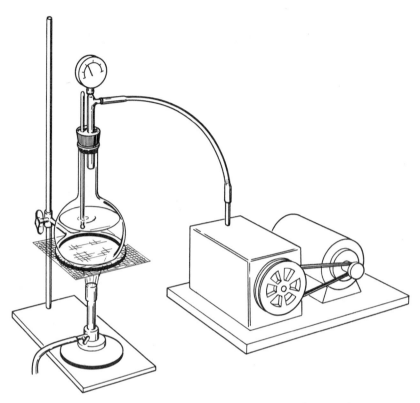

Figure IV-2. Experimental measurement of boiling point-temperature relationship.

(abbreviated to 14.7 *psi*) at sea level and only 11.9 *psi* at an altitude of one mile. The boiling temperature depends on the pressure of the gas (which includes air, of course) above the surface of the boiling liquid. As pressure decreases, liquids boil at lower temperatures (Figure IV-2).

If we boil a kettle of water (at 212°F), we know that we must continue to heat it for a long time before all the water boils away, that is, vaporizes and becomes a gas. This added heat, which does not increase the temperature of the water but is needed to turn the liquid at 212°F into a gas at 212°F, is called the latent heat of vaporization.

Suppose we take a pan of water which is just ready to boil at 212°F, cover it, connect it to a vacuum pump and quickly turn off the heat. As the pump decreases the pressure, the water will begin to boil very violently even though we are not heating it. This is because at lower pressure the water will boil and evaporate at a temperature below 212°F. But since we are not heating the water, and it takes heat to change from a liquid state to a gas (the latent heat of vaporization), where then is this heat coming from? The answer is really quite simple. The heat which is making part of the water evaporate is being taken from the remaining water. Remember that we described temperature as being related to the "average" internal energy of the molecule. This means that at any temperature, including the boiling point, some molecules will have more energy than others. Those with greater energy than average can escape from the liquid more easily, leaving the below-average, or cooler, molecules behind. Since only cooler molecules are left, the temperature drops. The above-average energy molecules, when they left the liquid, removed the latent heat of vaporization. Eventually, the remaining liquid cools to the new boiling temperature that goes with the pressure which we have reached. If we again decrease the pressure further, the boiling will again become violent and more liquid will evaporate. The remaining liquid will be cooler and will boil at still a lower pressure. We can plot a curve and find the temperature for boiling water at various pressures. (Figure IV-3). We can see from this curve that if we decrease the pressure to 0.34 psi, water will boil at room temperature (68°F or 20°C). This curve also explains much about why it is, for example, that clothes after washing, or being rained on, will

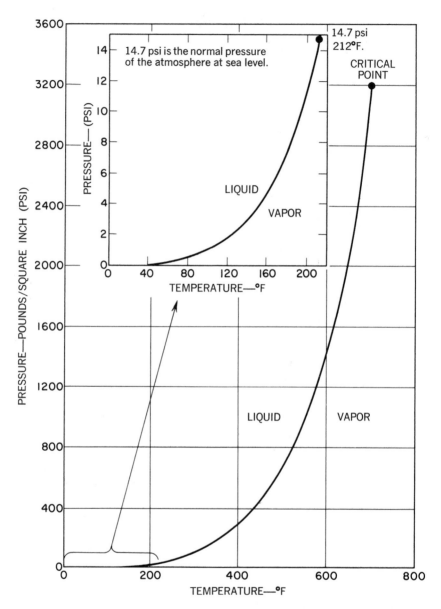

Figure IV-3. Vapor pressure curves for water.

dry, although water is not supposed to evaporate below 212°F at atmospheric pressure.

When more than one gas is present in a mixture (air, for example, which is made up mostly of nitrogen and oxygen), each gas acts much as it would if the other gases were not there. In fact, the pressure which we measure for the mixture is just about what we would get if we measured each gas by itself without the others and then added up the pressure of each. This concept is known as Dalton's Law. So while the atmospheric pressure as a whole may be 14.7 psi, the pressure of the water vapor alone at 68°F is 0.34 psi or less. If it is less than 0.34 psi, as it is when the humidity is low, the liquid, having a water vapor pressure of less than its boiling pressure at 68°F will evaporate. This pressure of the water vapor in a mixture is called its partial pressure, and it is this that determines the humidity.

We can determine the partial pressure on a given day from the relative humidity reported by the weatherman. The partial pressure is found by looking up the vapor pressure of water for the present temperature, and multiplying this by the relative humidity (in per cent divided by 100). For example, if the relative humidity is 50% and the temperature is 68°F, the partial pressure is 50% of 0.339 psi (found from Table IV-1) or 0.169 psi. We can see that the partial pressure can never exceed the vapor pressure at the same temperature since the relative humidity cannot exceed 100%. If the humidity is 100%, no water will evaporate and the clothes stay wet. When it becomes less than 100%, the water is below its boiling pressure and will start to evaporate, leaving the remaining water cooler.

If we think for a moment, this way of cooling by evaporating a liquid is not really new to us. In parts of the country

TABLE IV-1

VAPOR PRESSURE OF WATER
Points for Preceding Curve*

VAPOR PRESSURE POUNDS PER SQUARE INCH	TEMPERATURE °F
0.0886	32
0.126	41
0.173	50
0.245	59
0.339	68
1.07	104
2.89	140
6.87	176
14.7	212
28.8	248
52.4	284
89.6	320
145	356
225	392
336	428
3,206	705

* Francis Sears and Mark Zermanski, *University Physics,* Addison-Wesley Press, Inc., Cambridge, Massachusetts, 1950.

which have low humidity evaporation is used to provide air cooling. This is done by spraying a fine mist of water in front of a fan. You can build a similar makeshift air cooler by dipping a towel in water and hanging it in an open window. We know that some liquids, such as alcohol or ether, feel cold when poured on our skin. These are liquids which have very low partial pressures in our atmosphere. This causes them to evaporate rapidly. It is the latent heat drawn away from the unevaporated liquid that cools this remaining liquid and causes our skin to become very cold.

We have now learned that one way to get a lower temperature is to reduce the pressure over a boiling liquid, so that it gives up some of its internal energy by evaporation. If we want to lower the temperature significantly, we have to reduce the pressure greatly also (Figure IV-3), and evaporate a large amount of liquid. It is not so simple to do this, however, for we would need a large, expensive vacuum pump and a great deal of liquid. A simpler way is to use a liquid which boils at a temperature near that which we are trying to reach. The various chemicals, which we normally think of as gases, become liquids at low temperatures (Table II-1). As we have pointed out, the boiling temperature of these gases at atmospheric pressure differs. The problem of how to obtain these liquids, which normally boil well below room temperature, leads us to consider other means of producing refrigeration.

A slightly more complicated way of producing refrigeration by using this relationship between pressure and boiling point involves pressure rather than a vacuum. Suppose we take a gas, such as ammonia, which at room temperature is well above its boiling temperature ($-28°F$). Let us compress it and increase the pressure to about 43.9 psi above the normal pressure of the atmosphere; we find that it becomes a liquid. But like other gases, it becomes hotter. Touch the side of a tire pump after it has been used and you will see that this is so. We can keep the ammonia at room temperature, by either blowing air or pouring water over the chamber in which it is being compressed. In this way we cool the gas by again removing kinetic energy. When the pressure is later decreased, the ammonia will evaporate but the gas will now be cooler than when we started, because of the kinetic energy we removed with the cooling air or water.

However, there is a temperature for each gas above which no matter how high the pressure is raised, the gas will not become a liquid. This is called the critical point temperature for the gas. By looking at the upper end of the curve we drew, the critical value for water may be found (Figure IV-2). In the case of the cryogenic gases this critical temperature occurs below room temperature. For oxygen it is $-119°C$ ($154°K$) and for helium it is $-268°C$ ($5°K$). Since the cryogenic gases cannot be liquefied at room temperature no matter how high the pressure is raised, earlier researchers gave them the name of "permanent gases." Some new means of refrigeration had to be found before it became possible to liquefy these gases.

A very obvious way of lowering the temperature of a substance, and one which we have mentioned several times without giving it a name, makes use of a special device. This device is a "heat exchanger," and its principle is a simple one: if we put a warm substance near a cool substance, heat flows from the warmer substance and is transferred or "exchanged" with the cooler substance. To keep the two substances from mixing, in case they are different materials or at different pressures, the heat exchanger usually consists of metal tubes with one substance flowing through the inside of the tube and the other on the outside. A car radiator is a heat exchanger.

A very interesting principle, widely used in present-day refrigeration, was discovered in 1853 by two English physicists, James P. Joule and William Thomson (Thomson later became the Lord Kelvin after whom the temperature scale is named). The Joule-Thomson effect works like this: If a gas at high pressure is allowed to expand very rapidly by going through a nozzle to a region of low pressure, then the temperature of the gas may decrease. This only works at certain tem-

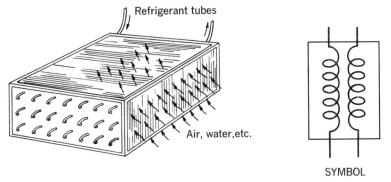

Figure IV-4. One form of heat exchanger.

peratures, and these temperatures are different for different gases. An example of this effect is the "snow" that sometimes forms when the gas is released from a CO_2 (carbon dioxide) cartridge or fire extinguisher. This happens because the gas has expanded and cooled so much that some of it becomes solid. The CO_2 "snow" which we see is at $-109°F$.

We have now looked at the ingredients we need to set up

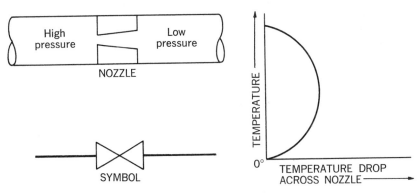

Figure IV-5. Joule-Thomson effect.

a practical refrigeration system, perhaps like the one that operates our own "icebox." Let us see if we can put one together (Figure IV-6).

We will start with the compressor, this takes the gas (the refrigerant), which enters at low pressure and at room temperature, and compresses it. As it is compressed, the gas gets hot, so we must next cool it back to near room temperature. In this way the heat leaves the refrigeration system. This

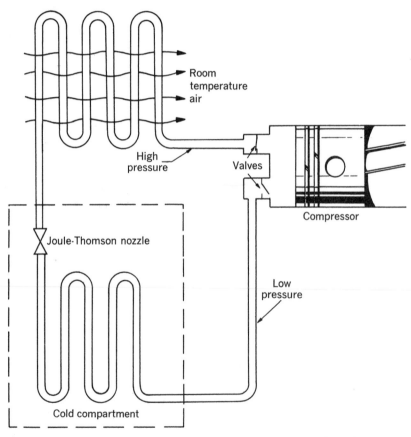

Figure IV-6. Refrigeration cycle in a simple household refrigerator.

cooling is achieved by passing air over the radiator coils of a heat exchanger, usually located on the back of the refrigerator. Since the gas is at high pressure, the cooling may cause it to become a liquid. This liquid or gas at room temperature and high pressure then goes to the part of the refrigerator which is to be cooled. Here it goes through a Joule-Thomson nozzle. The nozzle lowers the pressure, the temperature drops, and the fluid, if now liquid, returns to its gaseous state. As this cold gas circulates through tubes in the refrigerator, it absorbs heat from the walls of the tubes and from the inside of the refrigerator, thus providing us with the low-temperature compartment we seek for refrigerating. Now warm again, the low-pressure gas goes back to the compressor and the cycle is repeated.

This refrigeration cycle is very similar to the system set up by Linde many years ago to liquefy oxygen (Figure IV-7). But while our household refrigerator is a closed system and uses the same gas over and over again, Linde's machine used oxygen as the refrigerant gas, with more fed into it as the original oxygen became liquid. In his machine, when the oxygen expanded after going through the Joule-Thomson nozzle and became cooler, it was used in a heat exchanger to cool the gas which had not yet reached the nozzle. In this way, the gas reaching the nozzle kept getting colder, and in turn, the temperature of the low-pressure gas became still colder, until it finally started to become a liquid. Even then only part of the gas became liquid after going through the nozzle, and the rest, which was a very cold gas, was used to cool the incoming gas.

While it is so useful in the refrigeration systems we have just described, the Joule-Thomson effect does not work for all

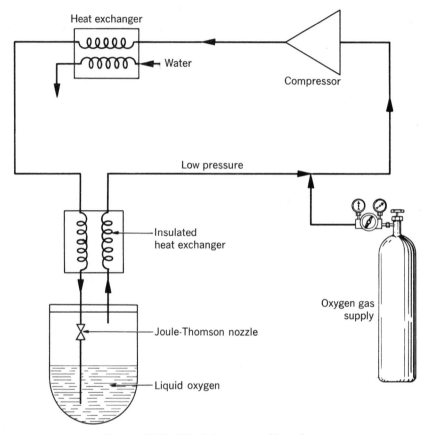

Figure IV-7. Linde's oxygen liquefier.

gases at all temperatures. As we said before, in some cases it actually makes the gas hotter. We cannot liquefy gases such as helium as simply as we just did the oxygen, for with helium the Joule-Thomson effect will not produce cooling until the helium is below 60°K. We already found out that helium's critical temperature (the point above which it will not liquefy) is only 5°K, so we cannot make it a liquid merely by compressing it. How do we cool the helium below this temperature of 60°K? One way is to use a heat engine.

In order to cool a substance, we must remove heat from it, the heat which we have called the internal energy of the sub-

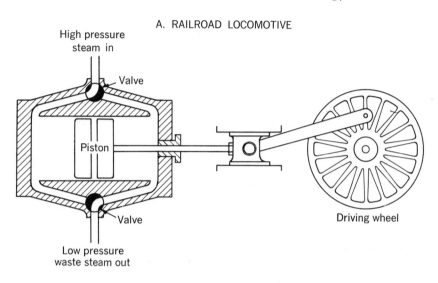

A. RAILROAD LOCOMOTIVE

High pressure steam in

Valve

Piston

Valve

Low pressure waste steam out

Driving wheel

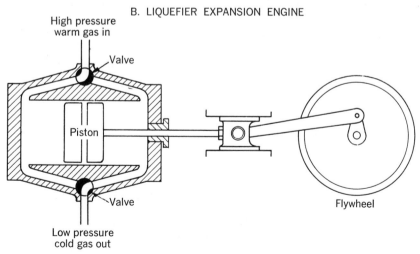

B. LIQUEFIER EXPANSION ENGINE

High pressure warm gas in

Valve

Piston

Valve

Low pressure cold gas out

Flywheel

Figure IV-8. Heat engines.

stance. Since this heat is energy, we can get rid of it by making the substance work, because work requires the expenditure of energy. So we make the substance work and operate a heat engine.

The heat engine may consist of a piston connected to a fly-wheel and a brake with some valves to make the piston go both ways (Figure IV-8). This is the way the old steam locomotives operated. The high-pressure gas (steam, in the locomotive) forces the piston down and this turns the flywheel (driver wheel in the locomotive) against the drag caused by the brake. In the case of the locomotive, the drag was due to the resistance of air to the moving train, plus the friction of all the wheels and axles on the cars. When the gas does work to overcome this drag, its energy as well as its temperature and pressure are lowered. In the locomotive, the steam was wasted after it had been used. But in the heat engine this output gas is saved, for this, after all, is the low temperature gas the heat engine has been used to produce. Another type of heat engine has a turbine connected to a brake or an electric generator, and uses the high-pressure gas to turn this turbine.

The refrigerator developed in 1950 by Professor Collins of M.I.T. became the first practical, widely used helium refrigerator. Many of the techniques we have described have been put together by Collins and provide a complete refrigeration system (Figure IV-9).

In this machine, the helium gas is processed through a four-stage compressor; that is, four compressors operate from the same motor, with the second compressor taking the gas from the output of the first compressor, compressing it further, and passing it on at higher pressure to the third compressor, and so on. At each stage, the heat added during compression is removed with water. The gas from the last compressor goes

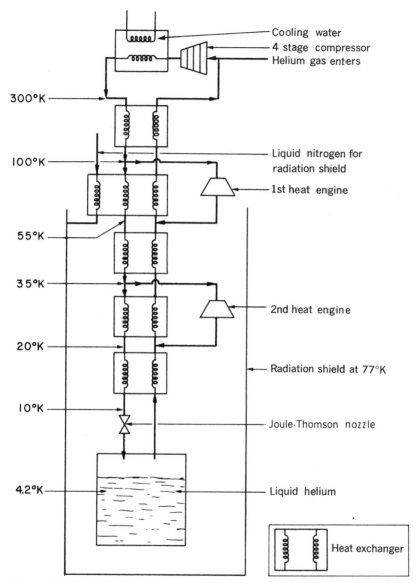

Figure IV-9. Schematic diagram of the ADL-Collins Helium Liquefier.

to the first heat engine. Here it does work in moving the piston and flywheel so its pressure and temperature are reduced. Part of the gas now goes down to the next heat engine to be cooled further; the rest goes back to the compressor, meanwhile passing through a heat exchanger where it helps to cool the gas following it to the first heat engine. In this way, the gas entering the first engine is cooled to 100°K. This first heat engine will finally bring the temperature of its outlet gas down to about 55°K. The same process which occurred with the first heat engine is repeated with the second heat engine. The temperature of the gas leaving this second engine is about 20°K. Again part of it goes back to cool the gas coming into the second heat engine (to 35°K) and also the gas coming into the first heat engine on its way back to the compressor. The rest of the cold gas coming out of the second engine is cold enough at 20°K for the Joule-Thomson process to work (remember, it works below 60°K for helium), so it is put through a nozzle where it is cooled further. Some finally becomes a liquid. The remaining gas is again used to cool the gas behind it, on its way back to the compressor to be reused.

The liquid helium refrigerators of the kind we have been discussing produced a temperature of 4.2°K (the boiling point of the helium). If we wish to go still lower in temperature, all we have to do is pump on the gas over the boiling helium and reduce its pressure. We mentioned this before, and it is just what Onnes did back in 1908 when he reached 0.7°K. But like this Dutch scientist, we find that it is very difficult to reduce the temperature of liquid helium much below 1°K because to do this, the pressure must be made very low indeed. We cannot use most of the other refrigeration techniques we have just talked about because the helium is already a liquid at this temperature and most of these methods

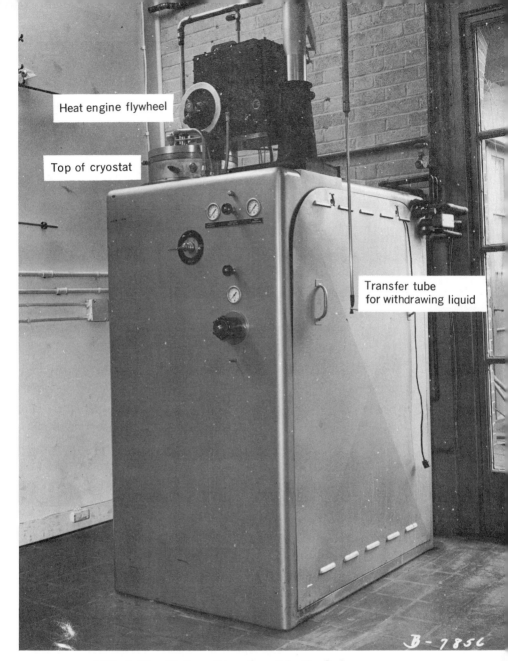

Heat engine flywheel

Top of cryostat

Transfer tube
for withdrawing liquid

ADL-Collins Helium Liquefier. (Martin Co.)

involve using a gas. So, if we are to reach a lower temperature, we must find a new way of doing it.

Refrigeration below 1°K is a field of science in itself with principles and methods quite different from those which we have considered so far. In trying to find a way to reach temperatures below 1°K, a method had to be found by which the temperature, or the internal energy, of a substance could be reduced to below 1°K. Liquid helium could then be cooled by being in contact with this substance. Because they could find no suitable gases and practically no liquids to reach this temperature, scientists started looking for effects in solids which they could use.

They described the phenomena they used as the *adiabatic demagnetization of paramagnetic salts*. This title and the combined effects it describes are complicated, but we can get a general understanding of what it means quite easily.

You are, of course, familiar with iron magnets. But do you know that there are other materials in addition to iron which can act as magnets, too? Among these are a number of ceramics and *salts*, which may possess various and differing properties. Certain of these materials act like a magnet when they are close to a strong magnetic field, but stop being magnetic as soon as this other field is removed. Materials with this property are said to be *paramagnetic*. A salt of this type gives us the *paramagnetic salt* part of our title.

Visualize the molecules which make up this paramagnetic salt acting like little bar magnets. We know that these molecules can move and turn, as we saw earlier when we considered what was meant by temperature. This motion, even at 1°K is still enough for these magnetic molecules to point in all different directions. Because they face every which way, there is no place on the whole block of paramagnetic salt that looks like the *north pole* (or *south pole*) of a magnet, and the whole does not act like a magnet at all. But as soon as we put the paramagnetic substance in a magnetic field, all of these little molecule magnets line up; the salt becomes magnetized and gives the substance north and south poles. Upon demagnetization the molecule magnets go back to pointing every which way; the north and south poles of the salt disappear, and it is no longer magnetic. If the magnetization and demagnetization is done without the salt gaining or losing energy, the process is *adiabatic*. Thus we have *adiabatic demagnetization of paramagnetic salts*. But we still have to learn why this produces refrigeration.

There are a number of forces and energies involved in a solid with these magnetic properties. As a result, we find that when the molecules are all lined up nicely in order they have less energy than when they are in disorder. The energy given up when the salt is being magnetized has to go somewhere, and it shows up by causing the temperature of the salt to increase. If it makes the temperature rise to above $1°K$, then we can cool the salt and remove this energy by using a helium supply cooled to $1°K$. Now if we demagnetize the salt after cooling to $1°K$, the molecules again tend to become disordered. But this requires energy, and the energy which went into heat when they became lined up has been removed. The only remaining source of energy is their own internal energy. In using some of this, of course, the temperature of the salt is decreased below $1°K$. If this cold salt is placed in contact with a second liquid helium container, which is ready to be cooled, it will absorb heat from the helium. This whole cycle of magnetizing, removing the heat with liquid helium, demagnetizing, and then cooling the second liquid helium, is repeated over and over, and each time a lower temperature is reached.

In our previous discussion of absolute zero we said that it was the lowest temperature predicted by theory, and that it could not be reached, even though scientists have come to within a millionth of a degree of it. We have considered some of the practical reasons why this point of absolute zero is very difficult to attain, but we have not yet actually proved its impossibility.

When designing a refrigerator (or any other machine), we are interested in how much energy we have to put into the

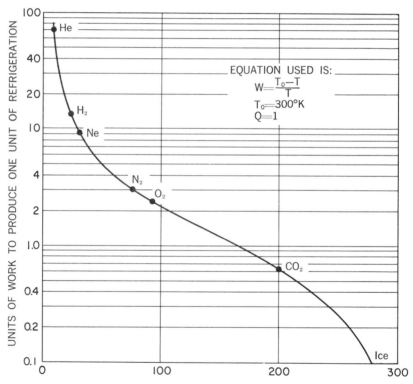

Figure IV-10. Work required to produce refrigeration. (Carnot)

machine for it to do the work we require. If it needs too much energy, like the ice machine built back around 1860 which used a pound of coal for every four pounds of ice it produced, then the machine is too costly to operate. In 1824, the French physicist Sadi Carnot found that, no matter how well the refrigerator was designed, a minimum value of energy was necessary to produce a given amount of cooling. Since that time, no one has ever reached the limits established by him, in spite of subsequent developments in heat engines and re-frigerators. One important aspect of Carnot's discovery is this: the amount of energy required to produce a given amount of cooling depends on the temperature of the space being refrig-erated and on the temperature of the space into which un-wanted heat from the system is being dumped. In a household refrigerator, this high-temperature dumping region is the room outside the refrigerator, and for convenience we will say its temperature is $300°K$ ($80°F$). If we call the amount of energy or work required W and the refrigeration produced Q, the high temperature T_0 and the refrigerated region tempera-ture T, we can write Carnot's equation (Figure IV-10) like this:

$$W/Q = \frac{T_0 - T}{T}$$

$$\frac{\text{Energy Required}}{\text{Refrigeration Produced}} =$$

$$\frac{\text{Warm Region Temperature} - \text{Cold Region Temperature}}{\text{Cold Region Temperature}}$$

We can see that as we approach absolute zero it takes more and more energy to produce the same amount of cooling, and that no matter how much energy we use we still cannot quite reach absolute zero. We can show this ourselves if we put numbers into the equation to find how much energy is required to produce one unit of refrigeration (Q = 1) with the warm region temperature at 300°K. If the cold temperature is 4°K, then W = (300 −4)/4 = 74 units; for the cold region temperature at 3°, then W = (300 −3)/3 = 99 units; for 2°K it is up to W = (300 −2)/2 = 149; for 1°K it is 299 units. Thus for a machine to produce 1°K takes 299 times as much energy as the amount of useful work, or refrigeration, that it performs. If you think this is a lot of energy for a small amount of refrigeration, try solving the equation for 0.001° and 0.000001° and you will see why absolute zero *cannot* be reached. You will find that no matter how much energy you use, the temperature still is not quite zero. It is much like trying to cut a piece of paper in half, then cutting one of these halves in half again, and then one of these in half a third time, and so on. The remaining piece keeps getting smaller, but it gets harder and harder to cut, and no matter how many times we cut it in half and how small it gets, we never reach the point where it is gone.

Chapter 5

THERMOMETRY, INSULATION, AND CRYOSTATS

TEMPERATURE cannot be described or measured as a physical effect. Under very restricted conditions, as we saw earlier, it may be identified with average molecular internal (or kinetic) energy. Generally we must resort to measuring temperature by relative methods; that is, by determining whether a substance is hotter or colder than another substance at a known temperature. Since we can never measure temperature directly, we must measure properties of matter which change with temperature. Fortunately, the many physical properties which do change with temperature provide many ways in which we can make thermometers. As we know, a thermometer is basically an instrument for measuring the change in a physical property which changes with temperature, with a scale that provides a way of relating this property change to the temperature change.

We know that mercury and alcohol expand when temperature increases. We measure the change in volume by putting one of these in a container, and marking the liquid level while the container is in an ice bath, and then in boiling water. Dividing the distance between the two marks into 100 even divisions, gives a Centigrade scale thermometer; 180 divisions gives a Fahrenheit thermometer. Of course, these thermometers cannot be used below the freezing temperature of the liquid which they contain, and thus are rarely applied to cryogenic work except for limited use at the warm end of the region. This liquid volumetric expansion type of thermometer was first built by a French physician named Jean Ray in 1631. Fahrenheit's later instrument was based on Ray's work.

There is another thermometer which uses a change in size, in which the change in length of a solid, usually a metal, is measured. This is called the *aneroid* thermometer. Since this change in length with temperature is very small, around 0.025% per degree, for example, it is very hard to measure. If the change can be made larger it can be measured more easily. By putting two different metal strips next to each other and joining them very tightly, this can be done. The whole assembly is called a bimetallic strip. Different metals expand different amounts for the same change in temperature. Metals in the bimetallic strip are chosen so that one has a large change with temperature while the other has a small change. Since they are joined together, the bimetallic strip bends as the two metals change length. This bend can be quite noticeable for even very small changes in lengths of the metals in the strip (Figure V-1A). One end of the bar is held rigidly, while the other can be connected to a needle on a dial to indicate temperature. Since the loose end of the bimetallic strip moves as the temperature changes, it can also be used to operate other devices, such as electrical switches, with changes in temperature. Bimetallic strips are used in thermostats for heating houses, in automatic toasters, and in many other situations where temperature must be controlled. However, they, too, are of limited value for cryogenic temperature measurements, since the length change with temperature becomes less and

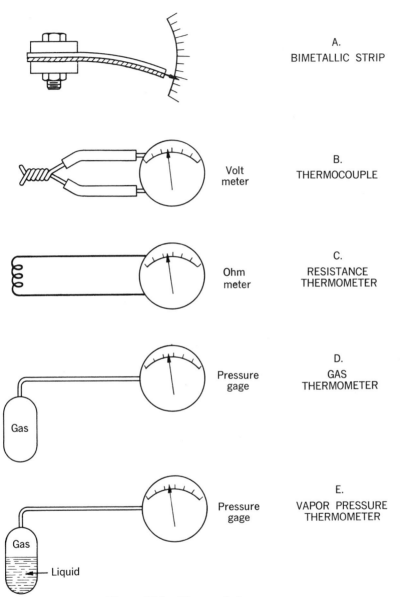

A.
BIMETALLIC STRIP

B.
THERMOCOUPLE

Volt
meter

C.
RESISTANCE
THERMOMETER

Ohm
meter

D.
GAS
THERMOMETER

Pressure
gage

Gas

E.
VAPOR PRESSURE
THERMOMETER

Pressure
gage

Gas

Liquid

Figure V-1. Types of thermometers.

less as the temperature gets near absolute zero. There is almost no change below about 60°K for most metals.

In 1821 Thomas Seebeck discovered that a current will be produced and flow through an electric circuit made up of two different metals, if their junctions are held at different temperatures. These junctions of dissimilar metals are called *thermocouples* (Figure V-1, B). A typical pair of junctions, using iron and constantan, will produce a voltage of 0.083 volts when their difference in temperature is 200°C. This voltage depends on the temperature difference, which is the reason the thermocouple can be used as a thermometer. Thermocouples are commonly used to measure temperature at liquid nitrogen temperature (77°K) in the upper end of the cryogenic region, and even sometimes at liquid helium temperature (4°K). The voltage produced by the low temperature thermoelectric junction becomes very small as we approach absolute zero. It becomes difficult to measure, limiting the usefulness of the thermocouple for very low temperature measurements.

Since these more common means of measuring temperature at normal temperatures are difficult to use, or fail to work at cryogenic temperatures, other lesser known techniques have to be used. These are the gas and vapor thermometers, the carbon resistance and platinum resistance thermometers, and the magnetic thermometers. Let us take a look at them.

The gas thermometer uses Gay-Lussac's Law, which we described earlier when we measured the value of absolute zero. In case you've forgotten, it states that the pressure of a constant amount and volume of gas is proportional to the temperature of the gas. We see that pressure readings can be used to tell temperature. To use the gas thermometer, a bulb containing a certain amount of gas is put in the low temperature region (Figure V-1, D). Usually helium gas is used, since it liquefies at the lowest temperature for any gas. The bulb is connected by a tube very small in diameter, to a pressure-gauge outside the cryostat, and this gauge is calibrated to indicate the temperature. This is one of the oldest forms of thermometers, and was thought of by Galileo around 1595. The instrument he used consisted of an inverted drinking glass containing gas, which was trapped as the glass was immersed in water (Figure V-2). Galileo measured the temperature by not-

Figure V-2. Galileo's thermometer.

ing the position of the water level inside the glass. However, since the water was exposed to atmospheric pressure, his thermometer was subject to errors caused by changes in barometric pressure.

The vapor pressure thermometer (Figure V-1, E) is in some ways similar to the gas thermometer (Figure V-1, D). It is very accurate, although operating only over a small temperature range. It uses a gas which becomes liquid near the temperature to be measured. Gas is put in a chamber which is located in the region where the temperature is to be measured. The chamber is connected by a tube to a pressure gauge or manometer at room temperature. As the gas becomes a liquid (at a temperature close to that being measured), some of it will start to condense. As it does so there will be less gas remaining in the chamber and the pressure will drop. Gas will continue to condense and the pressure in the chamber will continue to drop until the chamber reaches the temperature of the cryogenic region, and the gas is at its vapor pressure for that temperature. When the temperature goes up, the liquid will boil, and some will evaporate. This will increase the pressure of the gas in the chamber and raise the boiling temperature of the remaining liquid. This process will continue until the liquid is at the new higher temperature of the region being measured and the pressure has increased to correspond to the boiling point pressure for this temperature.

The carbon resistor thermometer uses the principle that carbon, a semiconductor, as are germanium and silicon, becomes an insulator at absolute zero. In doing this, its electrical resistance increases very rapidly below about 8°K. The ordinary carbon radio resistor works very well in this application. We will find out more about semiconductors at low temperature later on.

The platinum thermometer works much like the carbon resistance thermometer, except that the change in resistance with temperature is reversed. While the resistance of semiconductors like carbon goes up as the temperature approaches zero, the resistance of platinum, a metallic conductor, decreases. Measurement of the resistance tells the temperature.

But, like the thermocouple, it becomes difficult to read the resistance thermometer at very low temperatures, since the values become very small.

These cryogenic thermometers are limited to measurements above 1°K, much like the cryogenic refrigerators we first discussed. So we have to devise new methods of measuring temperature below this figure, just as we had to find new methods of producing the low temperature itself. Perhaps it is not really strange then that the use of magnetism, through which we obtained the very low temperature refrigeration in the first place, is also the means of measuring this temperature.

We remember that the amount of magnetism obtained with a paramagnetic salt increases as temperature decreases. If we measure this magnetism at several values of temperature between 4°K and 1°K (a region that other thermometers can measure also), we will be able to predict fairly accurately what the values of magnetism will be at temperatures below 1°K. Then the measurement of the magnetism of the salt in the temperature region below 1°K can be compared with the predicted values to find the temperature.

After producing and measuring this low temperature, we must find ways to retain it. A cryogenic fluid evaporates rapidly if it comes in contact with the comparatively hot objects in our natural environment. To prevent this outside heat from reaching the fluids, special containers called Cryostats must be built to hold them. To understand the principle on which their design is based, we must first understand the way heat moves from one place to another.

Heat moves from a higher temperature region to a colder region in three ways: by conduction, convection, or radiation (Figure V-3).

We can see the way conduction works if we put the end of

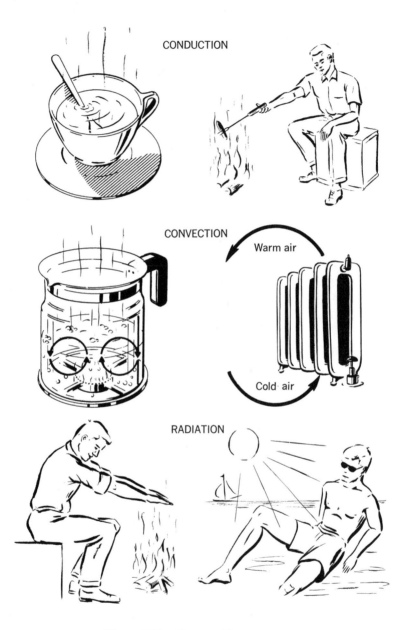

Figure V-3. Types of heat transfer.

a metal bar or a poker into the fire, or a silver spoon into a gravy dish. The metal itself doesn't move, yet the heat travels through it to our hand much like electricity through a wire; this causes the temperature of our hand to rise, even though it is far away from the source of the heat.

Convection is a little harder to understand. This process takes place in liquids and gases: those states of matter in which the atoms and molecules are free to move. What happens is this: first the atoms or molecules near the heat source are warmed up. This added energy changes some of their physical properties and they become more energetic, which causes the warm gas or liquid to expand. Because the warm part of the fluid is lighter (less dense), it is pushed away from the source of heat by heavier cold atoms and molecules. In moving to a colder region, it gives up most of the extra heat it just picked up. But as it cools it becomes heavy again, and so drifts back to the heat source. In turn it pushes away other warmer molecules until it reaches the heat source and is again reheated. By this *ferryboat* method the heat is effectively carried from the hot object to a cooler one. The radiators which heat many of our houses operate by this principle; the air, heated by the warm radiator, rises and spreads through the house, only to drop as it cools, returning to the radiator at floor level. This isn't so difficult to understand; but we can also do an experiment which will give us a visual picture and thus make even more clear what the transfer of heat by convection means. Heat some water in a pan. Just before it boils, add a drop of some bright vegetable coloring. The water is not still at all, and you will clearly see the "convection currents."

You yourself can feel heat transferred through the third method—by radiation—if you sit or stand before a roaring fire. The heat from the flames radiates out, and your face and everything exposed to the fire becomes very hot. Your back, on the other hand, turned away from the fire, is cold, because it is not exposed to the radiating warmth. We can also feel this effect by putting our hand near any warm object. Put your hand close to a hot iron. You can feel warmth radiating out. If you actually touched it, it might burn. So you see that heat is radiated from an object, even when the object is not so hot that it glows.

Now let us design an imaginary cryostat. If we think about each of these ways that heat can reach our low temperature material, the methods used in insulating and building such low temperature containers become quite clear. In the first case we considered, that of conduction, several things come to mind. If in holding the object, we grasp it at a point further away from the heat the part we touch is cooler. In addition we could protect our hand by covering it with cloth or plastic. This suggests that some materials are better at carrying heat than others, or, put the opposite way, some materials are better insulators than others. It also seems reasonable that a thick bar will carry more heat than a thin one, just as a large pipe can carry more water than a small one. From all this it appears that if we need a material extending from a high temperature region down to that of our cryogenic fluid, we would like it to reach over as long a distance as possible, be of material of poor thermal conductivity (a good insulator) and as thin as possible. With convection, we are concerned with the heat carried by molecules moving back and forth between the hot and cold regions. We want to make it difficult for them to move by putting material of poor thermal conductivity in their way, or by reducing the number of molecules available to carry the heat, such as we do by creating a vacuum. Finally, considering radiation, which we have not discussed in much detail, we can see that the larger the area of a heat source producing the radiation, the more heat it will give off. Most of us know that in the summertime a white or shiny surface reflects more heat than a black or dull one. The dull black surface will also radiate more heat than a shiny surface. So it is important to keep the warm surfaces facing the cold region as small and shiny as possible.

One of the older forms of insulated container still in wide use and one which illustrates the different forms of insulation, is the Dewar Flask. It was named after its inventor, James Dewar, who as we remember was also the first to liquefy hydrogen. The dewar flask (Figure V-4) is better known by its trade name of "Thermos." This flask consists of two glass bottles, one inside the other, joined together at the top but with a space left between them. This space, its walls coated with silver or some other shiny metal, is then emptied to a high vacuum. As a rule we use these containers to keep ordinary liquids such as tea or lemonade either hot or cold over long periods. But they also make very good cryogenic containers. For example, a quart container will hold liquid nitrogen (77°K) overnight. What is of real interest to us now is that in its design this bottle embodies so many of the desirable features for keeping heat from entering or leaving that we have just considered in designing our imaginary cryostat. The bottle is made of glass—a poor heat conductor; the glass is quite thin—also desirable since it keeps the amount of material through which heat can enter as small as possible. The inner bottle is connected to room temperature only at its top, providing the least possible amount of connection to high temperature. It also makes the longest possible path from the room temperature region to the liquid, particularly when the liquid level is lowered to near the bottom. Due to the vacuum

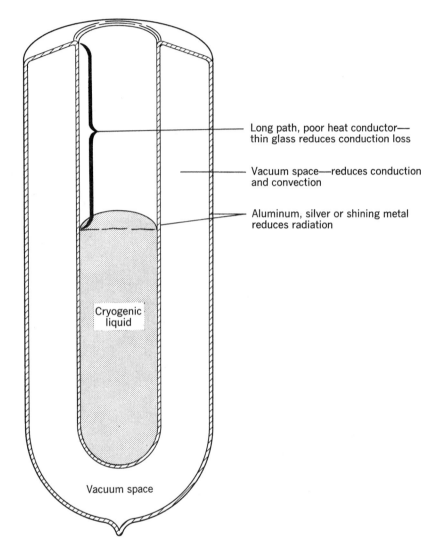

Long path, poor heat conductor—
thin glass reduces conduction loss

Vacuum space—reduces conduction
and convection

Aluminum, silver or shining metal
reduces radiation

Cryogenic
liquid

Vacuum space

Figure V-4. Ways in which insulation is used in the Dewar Flask.

between the two bottles, there are no gas molecules to support convection, and this form of heat transfer is eliminated. And the coating of the walls with a thin layer of silver or other shiny metal has reduced the amount of radiation.

The glass dewar is very important in the laboratory where small amounts of fluids are required, but it becomes impractical to use for large volumes of liquids. This has led to the development of specialized metal cryostats which involve a large number of construction methods and many different ways of providing insulation. While many of the smaller containers use a vacuum space surrounding the fluid, many others add some other insulating material as well. This material is often fiberglass or a substance called "Perlite," which looks like small powdery pieces of mica. Some cryostats use cotton or wool for insulation. A low-density plastic, such as expanded polystyrene used for making picnic boxes and ice buckets, not only acts as a good insulator without requiring the use of a vacuum, but can be molded to form a quite good cryostat by itself.

Recently, super-insulators have been developed. One of these shows well the principles we just discussed; it consists of alternate layers of thin plastic film coated with very thin layers of aluminum, about a millionth of an inch thick, and layers of fiberglass matting with the appearance of a white ink blotter. There are about forty layers to an inch of insulation thickness. The layers of aluminum reflect the heat, and the fiberglass acts as a very poor heat conductor. Between the two, very little heat penetrates the insulating layer.

The small storage cryostats used in industry and laboratories for holding liquid oxygen or nitrogen are as a rule double metal spheres that are separated by a vacuum space. These spheres are supported by tubes at the top. As cryostats become larger, it is no longer possible to suspend them from the top; insulated support pads or support wires in the vacuum space must also be used. As the containers become still larger, it becomes harder and more expensive to support them and to build them vacuum tight without leaks. As a result, large containers holding many thousands of gallons of fluids often do not use a vacuum at all but rely only on thick layers of insulation. These are often several feet thick. Serviceable cryostats, whether large or small, can be built in such a way that not more than a small percentage of their capacity evaporates in a day. This means that fluid left in a cryostat may take over a month to evaporate completely.

Helium and hydrogen cryostats are more complex than those built to hold nitrogen and oxygen. This is due largely to the low latent heat of vaporization of these liquids. This heat, which must be added to a certain quantity of liquid to turn it into gas, diminishes for gases which liquefy at very low temperatures. More care must be taken to insure against the entry of heat. The method usually employed is to surround the vacuum-jacketed fluid container with a second fluid chamber. This chamber is also vacuum jacketed and contains liquid nitrogen at 77°K. This gives us a cryostat within a cryostat. The inner cryostat, guarded by its low temperature casing, has only to guard its fluid against heat coming in from this 77°K region, instead of a room temperature region at about 300°K. Little use is made of hydrogen in research laboratories, for while it is less expensive than helium, it does not provide as low a temperature and requires care in handling to protect against explosion and fire. However, it is used in very large quantities for rocket propulsion. Through the use of adequate insulation and huge storage containers (Figure II-2), liquid nitrogen insulation is unnecessary and the amount of boil-off is kept low.

The size of the container plays an important part in determining its efficiency as a cryostat. If we consider that heat enters uniformly over the surface of a spherical cryostat having a 10-foot radius, which we will call R, we can agree that if

Liquid oxygen storage tanks at Cape Kennedy. This large tank is over 41 feet in diameter. Its size is evident from the height of the man at the right. (Chicago Bridge & Iron Co.)

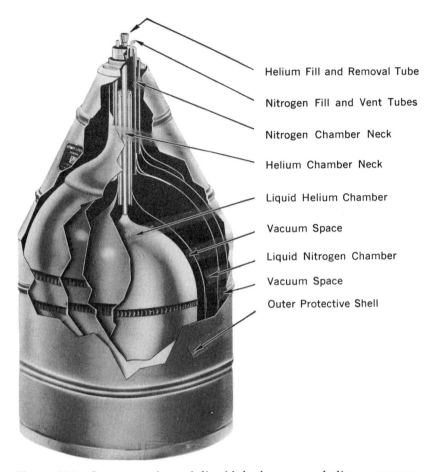

Helium Fill and Removal Tube

Nitrogen Fill and Vent Tubes

Nitrogen Chamber Neck

Helium Chamber Neck

Liquid Helium Chamber

Vacuum Space

Liquid Nitrogen Chamber

Vacuum Space

Outer Protective Shell

Figure V-5. Cut-away view of liquid hydrogen or helium cryostat. Cryostats of this type are used to store cryogenic fluids in the laboratory until they are used in the experimental cryostats. (Hoffman Laboratories, Inc.)

the radius of our cryostat is doubled, the surface (which equals $4 \pi R^2$) and the amount of heat entering will become four times as large. That is, if R^2 was 10^2 or 100 square feet, doubling R makes the new value of $R^2 = 20^2$ or 400 square feet; four times as large. The amount of liquid we can store in our cryostat is determined by the volume of the sphere; given by

4/3 πR_3. We find in doubling the radius from 10 to 20 feet that the volume increases from

$$\frac{4\pi}{3}(10)^3 \text{ or } \frac{4\pi}{3}(1,000) \text{ to } \frac{4\pi}{3}(20)^3 \text{ or } \frac{4\pi}{3}(8,000).$$

Doubling the radius has increased the cryostat capacity by eight times while it increased the heat input by only four times. Each liter, quart, or gallon of liquid now receives half as much heat as it would in the smaller cryostat. So, while the total evaporation from the larger cryostat is four times greater, the percentage is less, since it holds eight times as much liquid.

Helium cryostats are used in the laboratory for conducting many experiments and they vary widely in design. Some cryostats are so narrow they can be fitted between the poles of large magnets. Others have windows in the bottom to permit exposing the sample being studied to light or microwave energy. A cryostat for the study of frozen free radicals is shown on page 153.

Cryostats are frequently designed and built specifically for a certain type of experiment. They may be of either glass or metal construction, but one thing nearly all of them have in common is their insulation by vacuum and liquid nitrogen. The liquid helium cryopumping is often used in experimental cryostats. The liquid helium, of course, traps the particles that remain in the vacuum immediately surrounding the helium chamber. This reduces the heat entering the helium. In addition, the vacuum jacket around the liquid nitrogen chamber is often connected to the helium vacuum jacket. The helium cold trapping then improves the vacuum in both chambers and reduces the heat leaking into the nitrogen as well.

A syphon tube is used to withdraw cryogenic fluids from the laboratory type of storage containers. The tube is inserted in the neck of the cryostat until it almost touches the bottom of the fluid chamber. A collar placed over the tube where it enters the container seals the fluid chamber. The pressure of the gas above the liquid in the container is used to force the liquid up the tube and out of the storage cryostat. If the gas produced by the liquid evaporated in cooling the tube does not provide enough pressure, additional gas is forced into the chamber from a gas cylinder through a small tube in the collar. In the case of liquid helium, in which considerable gas is evaporated in cooling the tube, the extra gas is some-times collected and used to maintain the pressure later during the transfer. The transfer tubes for liquid helium and hydrogen are vacuum jacketed and this reduces evaporation.

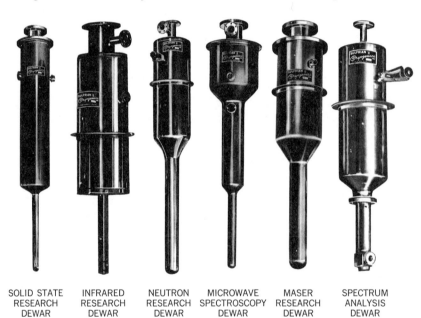

SOLID STATE	INFRARED	NEUTRON	MICROWAVE	MASER	SPECTRUM
RESEARCH	RESEARCH	RESEARCH	SPECTROSCOPY	RESEARCH	ANALYSIS
DEWAR	DEWAR	DEWAR	DEWAR	DEWAR	DEWAR

Figure V-6. Experimental liquid helium cryostats. (Sulfrian Cryogenics, Inc.)

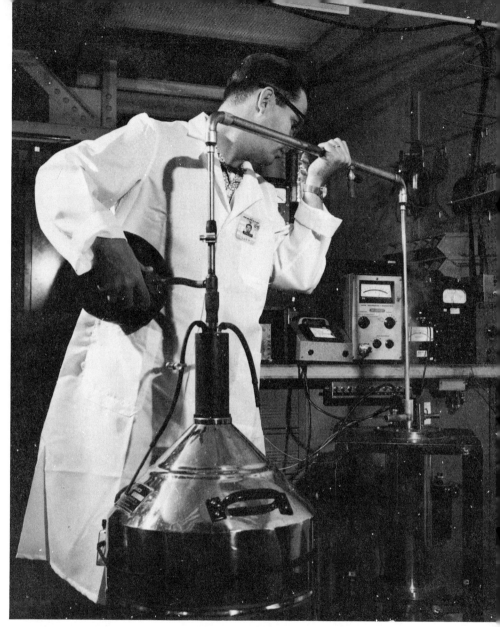

Transferring liquid helium in the laboratory. A vacuum jacketed syphon is used to transfer the liquid helium from the storage cryostat to the experimental cryostat. The gas evaporated in cooling the tube is collected in a basketball bladder and squeezed to maintain pressure during the transfer. (Martin Co.)

Chapter 6

CHANGES IN PROPERTIES OF MATTER AT LOW TEMPERATURES

MANY PEOPLE, before they understand cryogenics, are inclined to have the notion that as temperature goes down, everything becomes brittle and breaks. So, they figure, when absolute zero is reached, or nearly reached, all motion of atoms and electrons ceases and everything crumbles.

While this is not true, there are some things that do happen at low temperature that account for this line of thinking. We know of many substances, such as plastic and rubber, which become brittle at cold temperatures (though far above the cryogenic region). Because of this striking change, they are often cited in describing low temperature effects. The idea of mechanical breakdown at absolute zero was strengthened by the brittleness of steel at low temperature and the collapse of steel construction when used for cryogenic purposes.

The belief that everything fell apart at absolute zero was based on an early theory for the atomic composition of matter. It was the realization that this did not in fact occur which led scientists to a new and more accurate theory.

It is certainly true that steel becomes brittle at low temperature but it is also one of the few metals which does. It is an unfortunate coincidence that steel is one of the most common metals and is so widely used.

Most metals, including aluminum, copper, and even many types of stainless steel, do not become at all brittle at low temperature, and, in fact, even become much stronger. Today, care is taken to avoid using ordinary carbon steel for cryogenic devices and containers, and these can be built with little difficulty due to the good mechanical properties of other metals at low temperature.

Why is it that some metals are satisfactory for use at low temperature and others are not? Scientists have found that there is a direct relationship between the crystal structure of atoms making up a metal and its degree of brittleness at low temperature. Metals and alloys whose molecules have a *body-centered cubic structure* become brittle at low temperature. Steel is one of these. On the other hand, those whose molecules have a *face-centered cubic structure,* such as aluminum, copper, and nickel do not. It is the type of crystal structure of these metals that makes them valuable for cryogenic purposes. These crystal structures are quite easily understood.

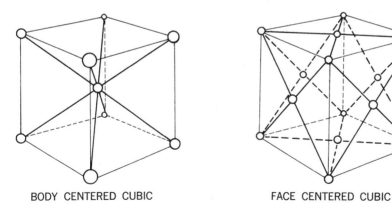

BODY CENTERED CUBIC   FACE CENTERED CUBIC

Figure VI-1. Crystal structures.

Visualize a cube, with an atom in each corner (Figure VI-1). If there is also a single atom in the center of the cube, we have the body-centered cubic structure we mentioned in relation to the structure of steel. If, instead, there are six additional atoms, one centered on each of the six surfaces of the cube, we have the face-centered cubic structure.

Most plastics become brittle at low temperature; others are usable but will crack or shatter if cooled too rapidly. This rate-of-cooling problem is often encountered with poor heat-conducting materials when they undergo rapid temperature change. Take for example, a glass container. Fill it with boiling water and put it in very cold water. When exposed to the cold, the outside of the container contracts; but because the untempered glass of our container is a poor heat conductor (or a good insulator) the inside of it is still warm and does not contract. This produces a strain and our glass will probably break. If we only gradually expose it to the cold, however, this is less likely to happen. So we have learned that if temperature is lowered slowly, the temperature is lowered uniformly throughout the material. This explains the fact that when handled with reasonable care, many plastics have been used successfully in building cryogenic equipment at liquid helium temperature. One plastic, Teflon, even remains flexible at helium temperature. This plastic and nylon are most often used to build cryogenic apparatus, but experimenters have made do with materials such as cellophane tape, airplane cement, and even chewing gum.

We have seen that when materials with poor thermal conductivities are cooled too rapidly, they will build up stresses or even shatter. There is another interesting effect associated with thermal conductivity at low temperature, and that is the change in the conductivity itself.

The thermal conductivity of a material means the ability of heat to pass through it. In an insulator, this heat transfer is done almost entirely by the atoms in the crystal lattice. Temperature affects the amount the atoms in the lattice vibrate. Motion of atoms in one part of the lattice causes the nearby atoms to vibrate also, and this in turn causes still others to move. In this way vibrations and heat travel through the material. In a metal, there is a second heat-conduction process. This is conduction of heat by electrons. In a metal, some of the electrons are free to move. When these electrons, which have a high energy since they are near the source of heat, move away through the metal, they carry this energy, or heat, with them. As a result, the metal has two means of conducting heat; by the lattice and by electrons. The ability to transfer heat by both these methods changes with temperature.

Materials with well-arranged, orderly crystal lattices are usually good heat conductors, but anything which interferes with this lattice makes them less good. This means that when the whole material is made up of only one crystal, it is a better conductor than when it is made up of many. Similarly, if the crystals contain impurities their conductivity will be lowered because the atoms of these impurities do not fit into the lattice perfectly. Pure materials with good crystalline structure usually become even better conductors as the temperature is reduced. In a metal crystal where thermal conductivity is due to electrons as well as the lattice, the combination of the improving lattice conductivity and deteriorating electron conductivity (as temperature decreases) causes the overall conductivity to reach a peak value and then deteriorate as the temperature nears absolute zero. A pure silver crystal, for example, will conduct as much as forty-five times better at about 60°K than at room temperature; while a pure copper crystal at about 15°K is twenty-five to thirty times as good as at room temperature. But copper that contains a small amount of other metals and is therefore impure, improves only about five times at 25°K and by the time the temperature is lowered to 15°K, its conductivity has returned to its room temperature value or less.

Because the thermal conductivity depends so much on the crystal make-up of the material, the metals used at low temperature where low heat conductivity is important are usually alloys, such as stainless steel. Since an alloy contains atoms of more than one element, its lattice is not as orderly as that of the pure metal. So it is not as good a conductor. The plastics, such as Teflon and nylon, have very disorganized crystal structures and this helps to make them very good insulators.

The heat conductivity by electrons is generally easier to predict than heat conducted by the lattice. The former normally decreases steadily as the temperature goes down, and is usually less important than lattice conductivity. However, the lattice conductivity sometimes decreases more rapidly with decreasing temperature, and this leaves electronic conduction as the main means of heat transfer in metals at very low temperature.

When we consider the way electrons conduct heat by moving through metal, we are getting close to the idea of electrical conductivity. The electrons which are free to carry heat in a metal are the same electrons which are free to move when an electrical voltage is applied across the metal. The ability of the electrons to move through the metal and transport heat provides the electronic contribution to the thermal conductivity. The ability of the electrons to pass through the metal when a voltage is applied is the measure of the electrical conductivity.

H. A. Lorentz, the Dutch physicist, found that dividing thermal conductivity by the electrical conductivity gave a number which changed very little for different metals, or for changes in temperature.

This relation between the thermal conductivity and the electrical conductivity creates a dilemma which makes it difficult to provide electrical wires to equipment operating at low temperatures. If the wire is a good electrical conductor, its accompanying good thermal conductivity allows heat to enter. While if it is a poor thermal conductor, it will also be a poor electrical conductor, and the electric current passing through this poor conductor will produce heat in overcoming its resistance.

Since electrical conductivity is defined as being equal to the reciprocal of resistivity (one divided by the resistivity), we will use the more familiar term of resistivity. Resistivity is simply the result which one obtains when the measured electrical resistance is divided by the length of the specimen and multiplied by the thickness. This makes resistivity a characteristic which depends only on the material from which the metal specimen is made, and what has been done in making it, and not on the size of the specimen.

As the temperature in a solid decreases, the atoms vibrate over a shorter distance about their position in the crystal lattice. Since they move less, their chance of colliding with an electron moving through the lattice decreases. The electrons do not bump into the atoms so much as the temperature decreases, and encounter less resistance to their passage through the metal. We have arrived at an important characteristic of metals: that their resistivity goes down as the temperature decreases. This, of course, also means that the electrical conductivity increases as the temperature is lowered.

The electrical properties of a semiconductor at low temperature are quite different from those of a metal. As its name implies, the properties of the semiconducting material fall between those of conductors and insulators. We are most familiar with it because it is used in transistors. The most common semiconducting elements are germanium and silicon. In its graphite form carbon is also a semiconductor, although not utilized for transistors. There are also many semiconducting alloys.

The semiconductor has fewer free electrons than the metal has. In fact, its outer electrons are attached to the atoms unless the temperature becomes high enough to allow them to escape, or an electrical voltage pulls them loose. At room temperature the semiconductor may have some electrons with enough energy to escape from their atoms and move through the material. If the temperature is lowered, the atoms and electrons will vibrate less and the resistivity will decrease, just as in a metal. But as the temperature is lowered further, more and more energy is needed to free electrons. As a result, fewer and fewer electrons will have enough energy to escape from their atoms. This means that there are fewer electrons to make up an electric current through the semiconductor and its resistance will increase as the temperature goes down. As we finally approach absolute zero, virtually no electrons have enough energy to escape from their atoms, and the resistivity of the semiconductor becomes very large. It is this property of a carbon radio resistor acting as a thermometer which is used to measure temperatures near absolute zero. The resistor, made of carbon, is a semiconductor. As it nears absolute zero, its resistance increases by a large amount for a small decrease in temperature, and it is this which makes it a sensitive temperature indicator. For example, a resistor which is 100 ohms at

300°K (room temperature) may decrease to 88 ohms at 77°K, due to the decrease in the motion of the atoms in the lattice. However, it will be about 150 ohms at 8°K, 1,000 ohms at 4.2°K, and 10,000 ohms at about 3°K.

The various cryogenic fluids which become gases long before they reach room temperature and their properties are interesting, particularly as we are unfamiliar with most of them (Table I-1).

Oxygen is at the high end of the cryogenic temperature scale, with a boiling temperature of 90.1°K. It has a light blue tint, but otherwise looks much like water. An unusual property of oxygen is that it is paramagnetic. This is the same property as that of the salts used for refrigeration below 1°K. We remember that when a material acts like a magnet when it is in a magnetic field, but not when this external field is removed, it is called paramagnetic. Because it is still chemically active, liquid oxygen can be dangerous to handle. Several serious explosions have resulted when liquid oxygen came in contact with lubricating oils. Because of this safety hazard, liquid nitrogen is substituted when possible for laboratory use. Liquid oxygen is encountered chiefly in the process of obtaining oxygen gas from air, as an oxidizer for rockets, and for storing and transporting oxygen for use as a gas.

Liquid nitrogen is the workhorse of the cryogenic fluids, used widely because it is inexpensive, chemically inert, and a good refrigerant. It is a clear colorless liquid which weighs about eight tenths as much as water. One quart of it will make almost seven hundred quarts of nitrogen gas at atmospheric pressure and room temperature.

As we know, both liquid nitrogen and liquid oxygen are normally obtained from air. (The dry atmosphere is made up of 21% oxygen, 78% nitrogen, and 1% argon and other gases.)

99

They can be separated by means of their different boiling temperatures. This can also work in reverse. Nitrogen boils at a lower temperature than oxygen, and a nitrogen-filled container, particularly one with a wide top, should never be left directly exposed to the atmosphere for very long. The air above the nitrogen will be cooled until some of the oxygen in the air itself becomes a liquid. It will then mix with the liquid nitrogen. As the nitrogen constantly evaporates and the oxygen condenses, the liquid in the container will eventually acquire the blue tint characteristic of liquid oxygen, and can even reach the point where a large part of the fluid is liquid oxygen. In this way, the formerly inert liquid takes on the dangerous properties of liquid oxygen and can become explosive.

Other rare gases which do not become liquid until cooled to well below room temperature (Table II-1) are: argon, krypton, xenon, and neon; they are obtained from the air by liquefication. All are chemically inert, and are not widely used in large quantities because they are expensive. Some of these inert gases have very desirable features. Xenon and krypton have boiling-point temperatures of 166.1°K and 120.1°K respectively, which are above the cryogenic range. Argon is the most common and its boiling point is 87.3°K. Neon boils at 27.1°K, which is close to that of liquid hydrogen at 20.3°K. Although neon is expensive, it is sometimes used as a substitute for the explosive hydrogen because it is inert. The very high latent heat of vaporization for this gas makes it a very good refrigerating fluid. These rare gases are all heavy when liquid. Their densities range from 1.2 (times as heavy as water) for neon to 3.1 for xenon.

Liquid hydrogen has the second lowest boiling temperature of any element, with only helium having a lower temperature.

Hydrogen is the lightest element, and therefore is the lightest liquid, weighing only about one fourteenth as much as water. We know that hydrogen is explosive and must be handled very carefully. For this reason, its boil-off gas is usually burned to prevent it from accumulating in dangerous amounts.

The liquid hydrogen molecule can take two forms, which are called orthohydrogen and parahydrogen; the two have slightly different properties. The hydrogen molecule consists of two hydrogen atoms, each with one electron. Electrons have a property referred to as *nuclear spin*. If the direction of nuclear spins of both electrons of the two atoms making up a hydrogen molecule are the same, it is said to be orthohydrogen. If they are in opposite directions, parahydrogen results. At room temperature, the gas is a mixture of about 75% orthohydrogen and 25% parahydrogen. When hydrogen is first liquefied, it has almost the same mixture as at room temperature, but after remaining a liquid for a while, it changes and becomes almost pure parahydrogen. The cryogenic engineer finds this a problem because the parahydrogen atoms have less energy than the orthohydrogen atoms. This means that when orthohydrogen changes into parahydrogen, it gives up energy in the form of heat, and so about one-third of the liquid hydrogen evaporates.

Since it ordinarily takes several days for the hydrogen to settle down to the right mixture, a cryostat filled at the liquefier would be partly empty by the time it gets to the user. To eliminate this evaporation, engineers try to speed up the change so that it will be completed before the liquid hydrogen leaves the liquefier. This is done through the use of a catalyst. A catalyst is a substance which makes a chemical reaction occur much more rapidly than it would if the catalyst were not present. The catalyst does not enter into the final

101

product of the reaction and therefore does not get used up. Several catalysts have been tried for this speed-up of the hydrogen change, among them hydrous ferric oxide. This is such a good catalyst that only about a liter (a little over a quart) is needed for a large liquefier. This is placed so that the liquid hydrogen trickles through it. Recently scientists have succeeded in converting hydrogen gas to parahydrogen at liquid nitrogen temperature. Since this means the conversion can be effected at a higher temperature than that of liquid hydrogen, it is more efficient.

In this chapter we have examined the properties of many solids and liquids at cryogenic temperatures. While most of these properties do differ from those with which we are familiar at room temperature, the changes are not too startling. In many cases they are pretty much what one would expect at low temperatures. There are several areas, though, in which very unusual effects do occur at cryogenic temperatures, and these must be examined separately and in more detail. These effects, occurring at the lowest level of the cryogenic temperature range, relate directly to liquid helium itself or to temperatures obtained through the use of liquid helium.

Chapter 7

LIQUID HELIUM AND SUPERFLUIDITY

HELIUM has some very strange and interesting properties, particularly as a liquid. In fact scientists find helium so interesting that in spite of its rarity and the difficulties encountered in liquefying it, they have studied it more than any other liquid except water.

Helium is the final gas on our temperature scale to become a liquid, and so it gives us the lowest-temperature liquid environment. In addition, helium is a safe cryogenic fluid to use. It does not burn like hydrogen, or support burning like oxygen. It cannot be made into a solid by pumping on it, no matter how low the temperature gets. Below 2.2°K properties of the liquid helium suddenly change, and it does things that do not happen in any other liquid at any temperature. We will go into these particular properties, especially one called "Superfluidity," later on.

Helium is the second lightest element. Only hydrogen is lighter. Because it is so light, when it is released it rises to the upper atmosphere. This means that in the atmosphere within our reach there is almost no helium. Helium forms very few chemical compounds, and is seldom found with other elements in liquids or solids. In fact, it is very hard to find at all.

Helium was detected first on the sun by analyzing the light during a solar eclipse in 1868. This light was recognized to be due to a new element by Joseph N. Lockyer three years later. Not until 1895 was it discovered on earth, when an English physicist, Sir William Ramsey, found it mixed with gas released from a mineral containing uranium, lead, and thorium. Helium was finally discovered in quantity on the earth in 1905—in Kansas, and mixed with natural gas. For many years helium was under governmental control because of its military value as a safe gas for dirigibles, but this is no longer an important use. Scientists outside the United States still find it hard to obtain today.

We have said that liquid helium does not become a solid at low temperature. Even at the *ground state*, or the lowest energy state that atoms have at absolute zero, these helium atoms still move too much and are too widely separated for the forces which would bind them together into a solid to be able to hold them in rigid positions. This can only be done through the use of pressure as well as low temperature, which makes the atoms stay close enough together for these forces to act. For helium to solidify requires a pressure of about 20 times that of the atmosphere, or about 280 psi, and a temperature of $3°K$, or less.

The helium atom has only two electrons. This means that the nucleus of the atom has two protons. The helium nucleus can have either one or two neutrons. If it has one neutron it is called the Helium-3 Isotope (1 neutron plus 2 protons equal three). Similarly, the helium atom with two neutrons in the nucleus is Helium-4 Isotope. The helium found in Kansas contains about ten million helium-4 atoms for every helium-3 atom. Recently the Atomic Energy Commission has separated and made artificially enough of the helium-3 atoms for scientists to make liquid helium-3 and study its properties. These were found to be quite different from those of helium-4.

Helium-3 and helium-4 boil at different temperatures and this helps to separate them; helium-4 becomes a liquid at 4.2°K, and helium-3 remains a gas until 3.2°K.

Like most cryogenic fluids, liquid helium has the appearance of water and is clear and colorless. However, it is only about one-tenth as heavy as water. It has a very low index of refraction, which makes it hard to tell where the liquid level is when looking through a transparent strip in a glass cryostat. Helium's very low latent heat of vaporization makes it very difficult to use and to keep, as we learned when discussing insulation. The latent heat is actually only about 5 calories per gram and since helium has a density of about 0.1, this is about 1/2 a calorie per cubic centimeter of liquid. It has been said that there is enough heat in an ordinary thumbtack at room temperature to evaporate a pint of liquid helium.

When the liquid helium-4 isotope is cooled below 2.19°K, some of its properties undergo a radical change. The helium enters a new state and becomes a new form of liquid. This is called helium-II to distinguish it from the ordinary helium-I, that is, the liquid helium above 2.19°K. These different types of helium are confusing to remember. The isotope 3 can only be in the helium-I state. Isotope 4 will be in the helium-I state if it is above 2.19°K and in the superfluid helium-II state if it is below 2.19°K (Table VII-1).

Suppose the temperature of liquid helium-4 at atmospheric pressure is lowered by pumping on it with a vacuum pump. While the pressure is being reduced, the liquid boils violently and its surface is very turbulent. But when the temperature of the liquid descends to 2.19°K, the boiling suddenly stops, there are no more bubbles, and the surface of the liquid

105

TABLE VII-1

STATES OF LIQUID HELIUM

Temperature	Helium Isotope	
	Helium-3 (Rare)	Helium-4 (Common)
Above 2.19°K	Helium-I	Helium-I
Below 2.19°K	Helium-I	Helium-II (Superfluid)

becomes as still as a millpond. This is our first indication of a property to be found only in the helium-II state: Superfluidity. In this superfluid state, the liquid does not appear to have any viscosity. In a viscous liquid the molecules try to adhere to each other, thus making the fluid less able to move. Molasses and motor oil, for example, become very viscous when cold. In helium-II, however, this viscosity does not seem to exist at all. The thermal conductivity appears to become very high and is at least 1,000 times as great as the conductivity of copper at room temperature. Actually, this is not thermal conductivity in the usual sense, for in helium-II, the cold molecules apparently pass between the warmer molecules without any friction (zero viscosity) and go to the warm region. Here they absorb heat and may go back to the helium-I state. This is why the boiling and turbulence in the liquid stops. The warm molecules can move to the liquid surface between the cooler molecules without disturbing them or producing bubbles. At the surface they evaporate and their heat of vaporization cools the remaining helium, just as in a normal liquid.

This apparent ability of the warm molecules to move between the cold molecules and vice versa suggested to scientists that they could consider helium-II to be made up of a mixture of two separate fluids: the superfluid, helium-II, and the ordinary fluid, helium-I. The proportion of helium-I to helium-II changes as the temperature is still further lowered below 2.19°K. The colder the fluid becomes, the more it behaves as if it were made up of more helium-II and less helium-I.

The ability of the superfluid to move without viscosity enables it to get in places where ordinary liquid helium would be blocked. Vacuum leaks which cannot even be found otherwise now become very serious. In fact, the superfluid will even leak through some glass, including Pyrex. This ability to go places where ordinary liquid helium cannot go produces some very interesting effects, as well as making things difficult for the experimenter.

A striking example of these strange properties of helium-II is shown by the helium fountain (Figure VII-1). To produce this effect we use a glass tube having a small nozzle at one end, and packed with a plug of fine emery or jeweler's rouge near the other end. We position the tube in the cryostat so that the nozzle is above the liquid helium level. While the plug prevents ordinary liquid helium from entering the tube, the liquid helium-II goes right on through. We now heat the inside of the tube directly above the plug with a resistance wire but only enough to raise the temperature above 2.19°K. This makes the superfluid return to the helium-I state. The ordinary helium-I cannot go back through the plug. The two forms of helium are now acting like the different gases we discussed in considering pressures and Dalton's Law in connection with humidity. The helium-II acts as if the helium-I were not present. As a result, more superfluid helium-II continues to come through the plug; warmed by the heat from the resistance, it becomes more helium-I. The pressure on this ordinary helium-I builds up and since the only way for it to escape from the tube is through the fine nozzle at the top, a stream of helium

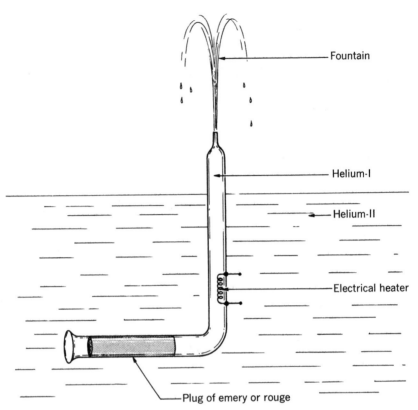

Figure VII-1. The fountain effect obtained with superfluid liquid helium.

squirts out through it. This gives the fountain-effect. This same idea could be used to make a pump with no moving parts, if ever there was the need for one at this temperature.

Another unusual property of the superfluid and one that is odd to watch is that of *creep*. A thin film of the liquid will climb up the walls of the container in apparent defiance of the laws of gravity. If we take an empty test tube, cooled to below 2.19°K, and push it partially below the surface of a container of helium-II, we can watch as the helium climbs the outside of the test tube wall, goes over the top, and then on down the inside wall until it fills the test tube and the liquid level is the same inside and out. But if we raise the test tube so that the bottom of it is above the outside liquid level, the creeping helium will now flow the other way, out of the test tube. It will climb back up the inside wall, go over the top, run down the outside wall, and drip off the bottom of the tube.

This problem of creep makes it difficult to design cryostats for helium-II, because the liquid will climb up the container walls. When it reaches a warmer region, it will evaporate. Naturally, this evaporation makes it more difficult to maintain the supply of liquid in the cryostat.

All these effects associated with superfluidity and helium-II occur only for the helium-4 isotope. As we have said, isotope 3 of liquid helium does not become a superfluid at $2.19°K$ and still has not become superfluid at $0.25°K$, the lowest temperature to which scientists have brought it. If theory is correct, it does not appear that this isotope would ever become a superfluid, no matter how low the temperature is reduced.

This difference in the two isotopes is carried much further in the theoretical explanations of the behavior of atoms. One theory for atomic behavior was proposed by Enrico Fermi and P. A. M. Dirac, and a second one by Jagadis Bose and Albert Einstein. Each theory is correct for certain atomic arrangements. One applies to the properties of the liquid helium-3 isotope and the other describes those of the liquid helium-4 isotope. They both agree with what is known as the quantum theory of matter. Helium-II provides one of the few ways in which this theory can be observed directly.

Helium-II is able to transmit temperature waves. No other material will do this. To understand what a temperature wave is let us consider how sound waves are transmitted. Sound is transmitted through a substance by means of a pressure wave. The sound wave, also called an acoustic wave, can be produced in many ways, by vibrating a tuning fork, for example, or the paper cone of a radio speaker. Regions of higher and lower pressures are produced, and these move away from the source like ripples moving after a stone is thrown into a quiet pool. After the sound wave passes a certain point, the pressure there goes back to what it was before. This sound wave can be reflected, and can also be used to operate a microphone. As a rule, temperature does not travel in this manner. We know that when heat is applied to one point in a uniform substance, this point will become hot, and some of the heat will flow out to the surrounding region, increasing its temperature, though a lesser amount. But when heat is no longer applied, the region of heat does not move away as the sound wave did. Instead the heat spreads out more, the temperature of the point where heat was first applied will decrease and eventually the whole substance is warmer.

However, this ordinary movement of heat does not hold true with helium-II. If heat is applied to a point in the superfluid liquid, by running electrical current through a heater, for example, a "wave" of heat will immediately start to flow away from the region containing the heater. If the current is turned off, the heat stops, and the region around the heater cools. However, the "wave" of heat continues to travel, just like the acoustic wave, causing a temperature increase while it is passing a point. The temperature drops again as soon as the wave passes the point. This is the manner in which pressure varies at a point when a sound wave passes by, and so this strange heat "wave" has been called *second sound*. While it has many properties like sound, it can be measured only with sensitive thermometers. A microphone will not measure it at all. The speed at which this heat wave or second sound travels has been measured and has been found to differ at different temperatures. At 1.8°K, it is 20 meters or about 65 feet per second. An ordinary sound wave or *first sound* travels at about ten times the speed of this heat wave in liquid helium at this temperature.

Let us look at second sound in a theoretical way and return to the superfluid liquid helium we experimented with before. We consider that the superfluid is made up of atoms of helium-I and atoms of helium-II; and we know too, that the higher its temperature, the higher the percentage of ordinary helium-I atoms. This suggests to us that as this second sound wave moves along, it actually is a region of more ordinary helium-I atoms and less helium-II atoms. When helium-I atoms flow into the region occupied by the wave, the helium-II atoms flow out, and this makes the temperature higher in the wave. The atoms can do this since they do not interfere with each other. The pressure does not change since as many helium-II atoms move out as there are helium-I atoms moving in. It is this lack of pressure change that explains why second sound cannot be heard with an ordinary acoustic microphone.

It is easy to see why liquid helium has been the subject of so much study. Not only are the physical phenomena discovered in the liquid helium enormously important in themselves, but through their study we have new knowledge of the behavior of atoms and molecules. Helium is the refrigerant that provides the low temperature environment for all experiments conducted below 5°K, and so it is often used as an instrument for conducting other research. Even Kammerlingh Onnes, after he succeeded in liquefying it in 1908, immediately began to use his liquid helium as a new laboratory tool for conducting low temperature research. Its use has contributed much to the advancement of science.

111

Chapter 8

SUPERCONDUCTIVITY

By NOW we are familiar with the name of Kammerlingh Onnes, and we know that the liquid helium he could produce permitted him to reach temperatures far below any which had ever been achieved before. In 1911 he was experimenting with metals at low temperature. Onnes was interested in finding out what happened to the electrical resistance of metals as they were cooled to near absolute zero. The atomic nature of matter was not well understood at that time, and a number of theories had been suggested. One step in determining which was correct, if any, was to find out what actually happened to the electrical resistance of metals near absolute zero.

We have already seen that the electrical and thermal resistance of metals decrease together as temperature decreases. The electrical resistance appears to be going to zero at absolute zero. But as Onnes experimented with platinum wires in liquid helium, he found this decrease did not continue at this lower temperature. The resistance curves became flat and retained resistance even as they approached absolute zero. He tried using different platinum wires and found that the curves became flat at different values of resistance. He decided this must be due to different amounts of impurities in the platinum, and to eliminate it he would have to use a purer metal.

Mercury is easy to obtain in a pure state. Since it evaporates at a very low temperature, we can purify it like water, by repeated distillations. Mercury is a liquid at room temperature. If you put some into a small glass tube and put it into the cold part of your refrigerator, it will solidify at $-38°C$.

When Onnes cooled his mercury wire, he discovered a very strange effect. While the resistance had been decreasing in the expected way as the temperature become lower, it suddenly

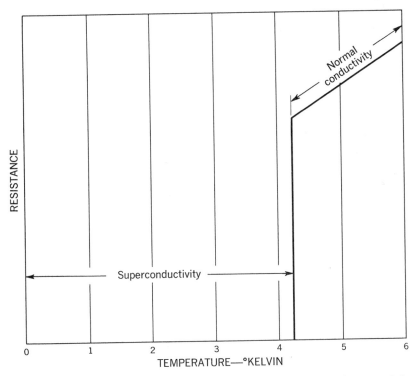

Figure VIII-1. The superconductivity of mercury, discovered by Onnes in 1911.

dropped to zero when the temperature reached 4.2°K. As best as he could tell, absolutely no electrical resistance remained in his mercury wire below this temperature. Since zero resistance means the metal is a perfect conductor, Onnes called this state superconductivity.

This experiment has been repeated many times with many different metals and alloys, and so far it has been found that at least twenty-two elements and hundreds of alloys become superconductive (Table VIII-l). These become superconduc-

TABLE VIII-1

SUPERCONDUCTIVE TRANSITION TEMPERATURES

ELEMENT	TRANSITION °K	ELEMENT	TRANSITION °K
Technetium	11.2	Aluminum	1.2
Niobium	8.0	Gallium	1.1
Lead	7.2	Rhenium	1.0
Vanadium	5.1	Zinc	0.91
Tantalum	4.4	Uranium	0.80
Lanthanum	4.37	Osmium	0.71
Mercury	4.2	Zirconium	0.70
Tin	3.7	Cadmium	0.56
Indium	3.3	Ruthenium	0.47
Thallium	2.4	Titanium	0.4
Thorium	1.4	Hafnium	0.35

tive at different temperatures: the highest for an element is 11°K for technetium, and 18.5°K for an alloy of niobium-tin. A few of these superconducting alloys are made up of two elements, neither of which will superconduct by itself. It is now thought that if metals can be made very pure, particularly if all the iron or nickel which they contain can be removed, that many more elements may be found to become superconductive. Other elements, bismuth and iron, are both superconductors

114

if they are allowed to evaporate and condense, thus forming a film on a support cooled to liquid helium temperature. Once this support is warmed, however, the crystal structure of the films changes, the superconductivity cannot be regained upon recooling and is lost forever.

Although Onnes had found no resistance when the mercury became superconductive, scientists could not be certain that the resistance was really zero. Perhaps it was just too small to be indicated by Onnes' instruments. It is not possible to measure something that has zero value. No matter how accurate the instruments are, the quantity you are trying to measure could be somewhere between zero and the smallest amount the instruments will measure.

Even though he could not expect to prove that metals had zero resistance when in the superconductive state, Professor Collins of M.I.T. undertook an experiment to show the resistance was less, at least, than could be measured by the most accurate method he could devise. Here is how he did it. First he cooled a lead ring to liquid helium temperature. Lead becomes superconductive at about 7.2°K, so the ring was well into the superconductive state. Collins then started a current flowing around the ring by using it as the output winding of a transformer. In this way he could get about 200 amperes to flow in a ring, looking just like a short circuit. A short circuit in our home would most likely blow out the fuse when the current reached 15 amperes. The fuse blows because the high current that flows through the resistance of the metal element in the fuse produces enough heat to melt the element. But if there is no resistance in the superconducting short circuit, no heat will be produced. This means that the metal will not melt, even though it is carrying many times the current which would blow out our household fuse. As a result, the current is not used up producing heat and melting the ring but will persist and continue to flow around and around the ring. Collins allowed the current to circulate around the super-conducting ring for two and a half years and then measured it to see how much the current had decreased. But even with instruments able to measure a change of one part in over 100 billion, he could not find any change in this current which had been stored by circulating around the ring for so long. We can show the rate at which the current would decrease in this way—

$$I = I_o \, \epsilon^{\frac{-Rt}{L}}$$

$$I = 200 \times 2.73^{\frac{-R \, \times \, 79,000,000}{L}}$$

I = Value of the current when time = t

I_o = Starting current = 200 amps

L = Inductance — depends only on the size of the ring and does not change

t = Time = 2½ years
= Almost 79 million seconds

R = Resistance

ϵ = Base of natural logarithm = 2.73

Since L does not change with time or with current, we can neglect its effect. The changes in the quantity $\epsilon^{\frac{-Rt}{L}}$ must depend on R and t. Since the current did not change in 2½ years $\epsilon^{\frac{-Rt}{L}}$ must be equal to 1. A mathematical term with an exponent only equals 1 when the exponent ($\frac{-Rt}{L}$ in this case) equals zero. We know t cannot be equal to zero since it is equal to 2½ years (or 79,000,000 seconds). This leaves us with the conclusion that R must be zero or very nearly so.

Shortly after he had discovered superconductivity, Kammerlingh Onnes tried to build a superconducting electromagnet, thinking it would be much more efficient than an ordinary electromagnet. The magnetic field produced by an electromagnet is the result of the electrons flowing through the wire. This means that a current must flow continuously. In an ordinary wire which has resistance, we must constantly supply energy in order for the current to flow. The current and magnet wire resistance will convert that energy into heat which we must remove. In large magnets, it is difficult to provide the energy and remove the heat. It is also a very wasteful device, since this energy served no useful purpose once the magnetic field was built up. In the superconducting magnet, however, there is no resistance, and so energy is only needed to produce the magnetic field. The current is able to circulate through the wire, keeping the field constant without requiring the addition of energy or removal of heat.

Unfortunately, Onnes ran into trouble. When he tried to produce a strong magnetic field, he found the metal stopped superconducting. Inadvertently in this way it was discovered that a magnetic field affects superconductivity. Not until several years later was this explained by the American scientist, Francis Silsbee, who gave rules for predicting how the destruction of superconductivity by an electrical current or magnetic field would occur. It was not until 1961 that scientists found a way of making superconductive magnets with very high magnetic fields that went far beyond Silsbee's predictions. We will go into these more recent discoveries later on.

Once the magnetic field effect was known, it became clear that the region of superconductivity is restricted to a location below a certain temperature and also below a certain value of magnetic field (Figure VIII-2).

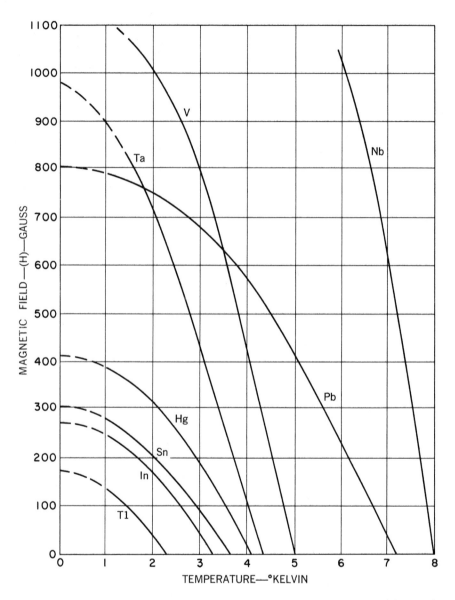

Figure VIII-2. Threshold magnetic fields versus temperature for various metals.

For many years scientists thought that superconductivity was the same as perfect electrical conductivity. Zero resistance is a basic requirement for perfect conductivity, and since many of the other properties of superconductors agree with theories for perfect conductivity, it seemed a sound conclusion. It was not until 1933 that Walther Meissner and R. Oschenfeld discovered that there was a difference. These two men found that while electrically the superconductor and the perfect conductor appear to be the same, their magnetic properties are quite different.

If a magnetic field is moved across an electrical conductor, it will cause or induce a current to flow in the conductor. This current in turn produces a magnetic field of its own. The current and its magnetic field are such that the field opposes the change in the original magnetic field. For example, if there was a magnetic field of 300 gauss (a unit of magnetic flux density) across an electrical conductor, and the field is reduced to 200 gauss, a current will be induced in the conductor. The magnetic field associated with this current would be 100 gauss, equal to the change in the original field. It would be in a direction opposite to the change and would add to the remaining 200 gauss to keep the magnetic field around the wire to 300 gauss.

If a perfect conductor had a magnetic field passing through it when it first became a perfect conductor, the field inside the metal would remain unchanged no matter what was done to the magnetic field outside. Why is this? Because if the magnetic field tried to change inside the perfect conductor, it would induce a current to flow inside the conductor. This would produce a magnetic field equal and opposite to the magnetic field change. Since the metal is a perfect conductor, this current and its magnetic field would persist indefinitely, as they did in Collins' experiment we described earlier. Therefore, since any change in magnetic field produces a current which persists indefinitely and since the new current has with it a magnetic field which exactly cancels out the original change in field, the total magnetic field will never change in value.

Meissner and Oschenfeld found that this was not completely true for the superconductor. In fact, they discovered a new effect, which is usually called the Meissner Effect. This states that as in the perfect conductor the magnetic field inside a superconductor not only does not change, but, in addition, is always zero. The magnetic field which might be passing through the metal before it becomes superconducting is pushed out when the metal becomes superconductive. This difference clearly showed that superconductivity was more than just perfect conductivity and caused scientists to search for an explanation. It is a useful addition to scientific knowledge and there are many situations where it can be applied.

One of the most impressive effects resulting from the fact that magnetic field cannot enter a superconductor is the floating sphere. The superconducting metal ball actually floats in space over a magnet coil. As the ball starts to fall into the coil, the magnetic field of the coil is squeezed further and further until it pushes back with enough force to hold up the sphere. The same effect can be achieved in reverse, by floating a small bar magnet over a superconducting metal cup. This idea of floating things in space by using superconductivity is used to eliminate the errors caused by friction in the ordinary supports of some gyroscopes.

The occurrence of superconductivity is often detected by a coil placed around the specimen. The specimen and some-

Superconducting sphere floating in a magnetic field. The lead covered ball is held up by the magnetic field produced in the coils. (A. D. Little Co.)

times even the cryostat are put in a weak magnetic field. When the specimen is cooled to the temperature at which superconductivity occurs, the magnetic field will be expelled from the superconductor. This will result in a change in the magnetic field through the coil closely surrounding the specimen, and cause a current to flow momentarily. Superconductivity can be detected by observing and measuring the current flow.

As scientists continue to study and experiment with superconductivity, they have been able to piece together more and more of their findings and arrive at more accurate theories. In 1935, two brothers, Heinz and Fritz London, found an explanation for the behavior of electrical current and magnetic fields on the surface of superconductors which fits in correctly with the equations for ordinary electrical conductors. Two scientists, C. J. Casimir and H. B. G. Gorter, worked on what happens to conductors when their temperature is lowered below the value at which they first become superconductive; this theory suggests that the superconductor has properties similar to those of superfluid helium. They considered the electrons to be like a fluid mixture, with some acting like a superfluid and some like an ordinary fluid. Alfred Pippard, an English scientist, measured the resistance of superconductors with high-frequency current and found that metals no longer superconduct when the frequency becomes too high. As scientists worked to explain the effects they found, theories concerning superconductivity became more and more accurate.

In 1957, three American physicists proposed the theory now believed the most nearly correct. This theory is called the Bardeen, Cooper, Schrieffer, or BCS Theory, after its originators. John Bardeen, who is given a large amount of the credit, had already won a Nobel prize in physics for work leading to the transistor. The Russian scientist, N. Bogoljubov, expanded the BCS theory, but while he cleared up some shortcomings, he also uncovered additional problems.

The BCS theory is extremely complicated and we cannot hope to develop an understanding of it here. We can follow a few of its major points, however, and these may give us some idea of what happens in a superconductor.

When we considered electrical conductivity in ordinary metals we learned that we would achieve perfect conductivity if the electrons could move, with no collisions, through the atoms forming the crystal lattice of the metal. Scientists now know superconductivity is not the same as perfect conductivity because of the different magnetic properties. In the superconductor, there could be no magnetic field and any which had been there before the material became superconductive was pushed out. In addition, there were a number of practical ways to show the error in the explanation which said the electrons miss the atoms and thus encounter no resistance. The lattice would have to be perfect, with no impurity atoms or defects where two different crystals came together. In admitting that the electrons could not be expected to avoid hitting all the atoms, we have implied that they must hit some, but that in spite of these collisions there still is no resistance. This implication was made much stronger when it was found that the superconductive characteristics of two different isotopes of the same metal were different. Since isotopes have the same number of protons in the nucleus and the same number of electrons, they are identical electrically. But when the superconductive properties of different isotopes were found to be different, it was realized that this change must be due to something other than the electrical properties of the atoms. Isotopes differ only in the number of neutrons the atom has in the nucleus. Since the neutron has no electrical charge, the different superconductive properties of isotopes must be due to the different mass, or size of the atoms. This would only be noticeable if the electron hit the atom.

The BCS theory suggested that electrons travel in pairs through the superconductor. When one of the electrons collides with an atom, the energy it loses, which would ordinarily go into heat, instead causes the atom and the lattice to vibrate more than before. This extra vibration, which is called a phonon, carries the energy from the collision through the crystal lattice to the other electron of the pair. It can do this even when these two electrons are separated by many atoms of the lattice. When this phonon vibration carries the energy to the second electron, it can use this extra energy to help pull the first electron along. As a result, the energy involved in the collision has gone from one electron to the other and not into heat.

The electron pair continue on in this way through collision after collision without giving up energy. As long as the electrons can stay together in pairs the metal acts like a superconductor. To do this they have to stay less than a certain distance apart. If they separate by more than this *coherence length,* they are no longer paired and lose their superconductive properties. Either a strong magnetic field, or an increase in temperature can provide forces which will break up these electron pairs and so destroy superconductivity. This explains why superconductivity occurs only at low temperature and in weak magnetic fields.

In this quick look at the theory of superconductivity, we have caught only a glimpse of an area of science in which man is daily uncovering some of the secrets of nature. The *exact* nature of superconductivity is still not known. The BCS theory, which does much to explain it, was the result of improving on many earlier ideas. Scientists such as Bogoljubov have corrected it, added to it, and further refined it. Each time there is a new discovery, the theory must be re-evaluated to see if it will explain the new effect. The theory also predicts effects which the scientist then looks for by making experiments. The evolution of this theory shows particularly clearly that while much scientific knowledge is gained by the brilliant independent discoveries of the Newtons, Maxwells, and Einsteins, much more is the result of long and continuous work on the part of many scientists, each adding to the store of knowledge until enough evidence has been collected to provide a clear picture of the workings of some part of nature.

Chapter 9

CRYOGENIC ELECTRONICS

CRYOGENICS, and especially the phenomena of superconductivity have shown great promise of application to electronics. Not only can totally new devices be made which will operate only at these low temperatures, but many ordinary electronic devices will perform much better when cooled to either liquid nitrogen or liquid helium temperatures.

The idea of applying cryogenics to electronics is certainly not new. Back in 1914 Kammerlingh Onnes, for example, tried to build a superconducting electromagnet. Although it was not successful in producing the large magnetic fields he hoped for, it and many other ideas were suggested for cryogenic electronics.

The main factor preventing the application of cryogenics to electronic devices at that time remains the major obstruction to their use today. This is the refrigeration problem. Until 1923 Onnes' liquefier was the only one in existence. For many years, actually up until 1950, when the ADL-Collins Liquefier was first made in reasonable quantities, liquid helium was a laboratory curiosity that was very difficult to obtain. The physicist who needed it often had to build his own liquefier. If he managed to design and build a helium liquefier and produce some liquid he had done a significant piece of work, even before he began his low temperature experiments. Outside the laboratory there was no way to keep the electronic devices cold enough to operate and so they had no practical application. Researchers were concentrating on those basic physical effects which contributed to our knowledge of low temperature phenomena, and rightly they left the electronic devices for such a time as they could be effectively utilized.

Some of the first cryogenic electronic research was done by Dr. Donald Andrews at the Johns Hopkins University during World War II. He was trying to build a liquid hydrogen cooled superconductive device to detect radio waves. To get the highest possible transition temperature, he used the niobium-tin alloy, which superconducts at $18.5°K$. This allowed Andrews to use liquid hydrogen at reduced pressure as the refrigerant. His device, called a bolometer, converted radio waves to heat and measured the increase in resistance which occurred as the niobium-tin became warmer and lost superconductivity. The temperature had to be carefully controlled so that the niobium-tin was part way between being a superconductor and a normal conductor. When a radio wave struck the bolometer, its energy was changed into heat. This heat increased the resistance which remained in the niobium-tin and made it more like a normal conductor. The transition between superconductivity and normal conductivity occurs over a very narrow temperature change, so it required only a very weak radio signal to produce a resistance change large enough to be measured. Although the superconducting bolometer worked, unfortunately many problems prevented its use. Since Andrews' work, it has become much easier to obtain and keep liquid helium. Work has begun again in several laboratories to develop superconducting bolometers made of tin which would operate at about 3.7°K.

The real start of cryogenic electronics can probably be said to have occurred when Dudley Buck of M.I.T. published a paper in 1954 which described the *cryotron* and how it could be used to replace transistors and vacuum tubes in electronic computers. While the idea of the cryotron was not new, it had previously been put aside as impractical because there was no means of providing the necessary refrigeration.

Buck was making a study of the various effects in nature which could be used to build electronic computers. By 1954, the vacuum tube used in the past was being replaced by the transistor, which offered many advantages. But Buck and other scientists were looking further, for ways of building electronic systems with greater capabilities, in smaller space, with more reliability and at lower cost. They were not satisfied with the step forward provided by the new transistor.

In considering reasons for the rejection of superconductivity by earlier engineers in designing electronic computers, Buck found the situation had changed. The difference was due to improvements made in helium refrigerators by Collins, also of M.I.T. Apparently liquid helium would soon be easily obtained. So the use of superconductivity was reinvestigated, the principle of cryotron was reconsidered and its name bestowed.

The cryotron is an electronic device doing much the same task as the vacuum tube or the transistor. It operates on the principle that its electrical resistance will be considerably different in the *normal* state than in the superconductive state. This means it can perform in much the same way as a switch with a low-resistance "on" position and a high-resistance "off" position. It uses the fact that a magnetic field will destroy

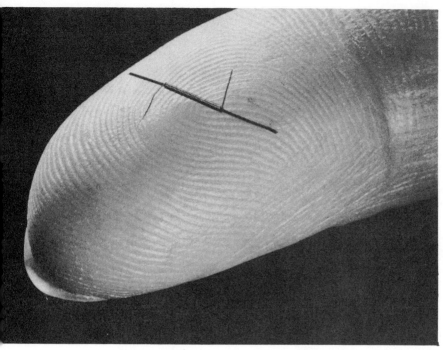

Wire wound cryotron. (A. D. Little Co.)

superconductivity as the means for switching from one resistance value to the other.

The original cryotron developed by Buck consisted of a straight piece of tantalum wire 0.010 inches in diameter, called the *gate*, around which was wrapped a single layer of 0.002 inch niobium wire, called the *control*, thus forming a magnet coil about ½ inch long.

The way in which the cryotron operates is really quite simple (Figure IX-2). Tantalum becomes superconducting in zero magnetic field at 4.4°K. As a result, it is just barely superconductive when immersed in liquid helium at 4.2°K.

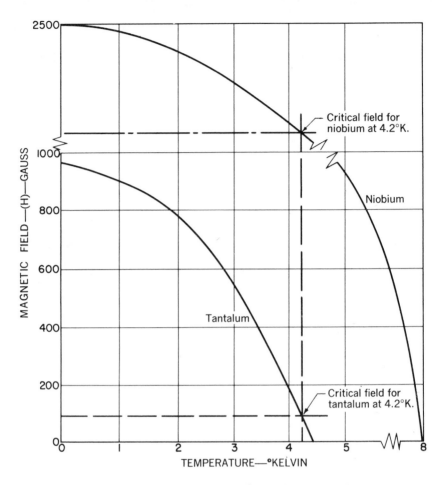

Figure IX-1. Critical field curves for the niobium-tantalum cryotron.

This means that only a weak magnetic field is needed to cause the tantalum to change back into its normal conductivity state. The niobium control coil which superconducts at about 8°K is well within its superconducting region at 4.2°K. As a result, the magnetic field produced by a current flowing through the niobium control wire will cause the tantalum gate wire to stop being superconductive, while the superconductivity of the niobium control is not affected.

Many electronic circuits were proposed to enable the cryotron to perform functions in electronic computers. One of the simpler circuits, which illustrates how the cryotron is used, is the Bistable Multivibrator Circuit. This circuit is commonly called the "flip-flop." The flip-flop is able to store or remember a single piece of information. It can remember a "yes" or "no," a "one" or "zero," an "on" or "off," or any other such item of information that the designer selects.

The flip-flop works in much the same manner as a traffic light. Information as to which cars must stop is stored by

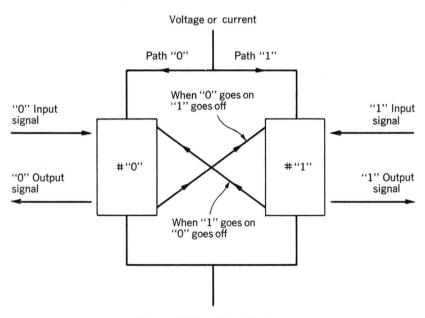

Figure IX-2. The flip-flop.

means of light bulbs. When the main-street light says "go" the side-street light says "stop." When a new signal is received from the control box, the conditions will change, and the main-street light changes to "stop," the side street to "go." This is the way the electronic flip-flop functions (Figure IX-2).

There are two paths the signal can flow through, each containing an active component, such as a vacuum tube, a transistor, or even a cryotron. The resistance of one active component will be low and the other will be high. They cannot both be high or low at the same time, just as a traffic light cannot be green in both directions. Which component is high and which is low depends on the last signal received by the flip-flop. If No. 0 has a low resistance and No. 1 is high, the circuit is said to be remembering a "yes," "one," or "on." If No. 0 is high and No. 1 is low, the circuit is remembering a "no," "zero," or "off." By looking at Figure IX-3 and Table IV-1, page 134, we can see how this is done with the cryotrons.

Before the current is turned on, neither cryotron gate is resistive, and neither control winding is producing any magnetic field. This is Step 1. Suppose an electric current enters the circuit at the top; since both paths cannot be built perfectly the same, more current starts to flow down path A than B. This current flows through the tantalum gate of the "1" cryotron, but does not produce enough magnetic field to destroy its superconductivity. It also flows through the niobium "control" wire of the "0" cryotron. In this coil it produces enough magnetic field to make the tantalum "gate" of cryotron 0 resistive, but not enough to destroy superconductivity in the niobium control wire. Since neither the tantalum gate of cryotron 1 nor the niobium control of cryotron 0 have any resistance, there is no electrical resistance in Path A. The gate element of cryotron 0 in Path B is resistive because of the

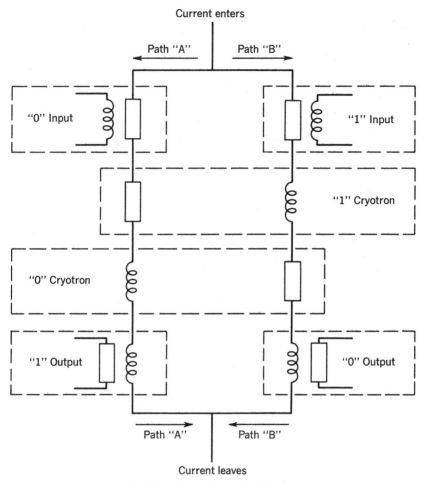

Figure IX-3. Cryotron flip-flop.

current in the coil. This means that there is a resistance in Path B. With no resistance in Path A and resistance in Path B, all the current will flow in Path A. A "1" or "yes" or "on" is stored. This completes Step 2.

STEP	Path "A"					Path "B"					Signal stored
	Current	Gate Input "1"	Gate "1" Cryotron	Control "0" Cryotron	Control "1" Output	Current	Gate "1" Input	Control "1" Cryotron	Gate "0" Cryotron	Control "0" Output	
1. Current is off	0	0	0	0	0	0	0	0	0	0	None
2. Current is turned on, and "1" is stored ("0" cryotron gate becomes resistive)	I	0	0	H	H	0	0	0	R	0	1
3. Signal current is put in "0" input cryotron control		R									1
a. Current divides	$\frac{1}{2}$					$\frac{1}{2}$					Changing
b. "1" and "0" cryotrons act alike			R					H		R	
c. 2 resistances in path "A" cause current to go through path "B"	0					I					0
d. Switching is complete											
e. Current is removed from "0" input cryotron											
4. State has been changed, "0" is stored.											0

Cryotron "0" switches

Cryotron "1" switches

Not used to switch from "1" to "0"

I—means current is flowing.
R—means gate is resistive, superconductivity has been lost
H—means magnetic field is present, due to current flowing through the control winding.
0—indicates absence of either R, H, or I.

Shading indicates I, R or H.
No shading indicates absence of I, R or H.

TABLE IX-1   OPERATION OF THE
CRYOTRON FLIP-FLOP

Suppose now that we want to change the information stored in the cryotron flip-flop from a 1 to a 0. This brings us to Step 3. We put a current through the 0 input cryotron control coil in Path A. This makes the gate restive and puts a resistive gate in each path. The current coming in the top of the circuit momentarily divides evenly between the two paths. The selected value of this current is such that when it divides into two paths, neither has enough current in the control coils to make resistive the gates in the other paths. Now the gate of cryotron 0 is no longer resistive. Since the gate of the 0 input cryotron remains resistive as long as we are putting a current through its control coil, there is a resistance in Path A but none in Path B. The current flowing at the top of the circuit will now all flow through Path B. As soon as this happens, the current flows through the control coil of cryotron 1, and this makes the gate of cryotron 1 resistive. Since this gives a second resistance in Path A, it means that we can now stop the input current through the 0 input cryotron control (Step 4), and the flip-flop current will remain flowing through Path B. The flip-flop is now remembering a "0" or "no" or "off." The output cryotrons at the bottom of the flip-flop circuit can be used as the input coils to another circuit or give a way of telling what information is stored in the flip-flop.

When these circuits were built, it was found that the wire cryotrons had a number of disadvantages. They were hard to make and to handle, they had to be welded together, and they operated much more slowly than transistors or vacuum tubes.

Many of these problems were overcome by thin film technology. Instead of making each cryotron from wire and then connecting many of them together to form an electronic circuit, the circuit was made by spraying thin layers of metals and insulators through a stencil onto a flat sheet of insulating material, usually glass. This sounds much easier to do than it actually is, for several years of work were required to develop the techniques for making the thin film superconductive circuits. Much is still to be done.

The spraying of the metal and insulators to make up the superconductive circuitry is achieved by heating them in a vacuum chamber until they evaporate and form the spray. The vacuum keeps the metal vapor from reacting chemically with air, or from being scattered by bumping into air molecules. The stencils are usually made of metal, and several are used to make up a circuit, each one consisting of a single layer of one kind of metal or of an insulator. The several masks required to build a circuit all have to be used, and the circuit must be completed, without opening the vacuum chamber.

This thin film vacuum deposition technique produced a radical change in the appearance and performance of the cryotron circuitry. While niobium and tantalum had been used in the wire-wound cryotron to permit the device to operate at the normal boiling-point temperature of helium ($4.2°K$), neither of these materials could be vacuum deposited easily. They both have very high melting and boiling temperatures, and will not superconduct if they contain even very small amounts of impurities resulting, for example, from the remaining atmosphere in the vacuum chamber. Lead and tin are metals well suited for vacuum deposition, because they evaporate at low temperatures and are not as sensitive to impurities as niobium and tantalum. However, they also superconduct at lower temperatures, $3.7°K$ for tin and $7°K$ for lead. This means that the pressure of the vapor over the helium has to be reduced and the liquid temperature lowered

136

to slightly below 3.7°K before the lead-tin cryotron can operate. While this is a minor problem, it does make it more difficult for the research scientist who can no longer just "dunk" his device in the liquid helium.

When making thin film cryotrons, it is quite difficult to form the equivalent of a coil such as the niobium control coil used in the wire-wound cryotron. Scientists quickly realized

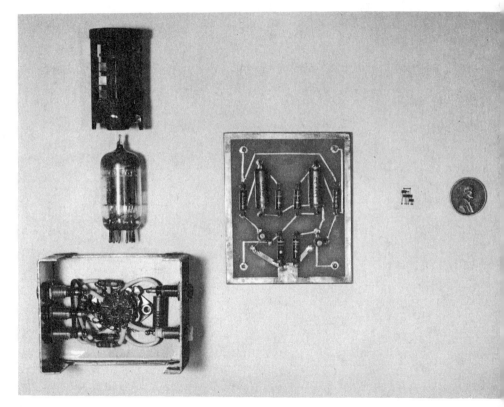

Comparison of vacuum tube, left, transistor, center, and cryotron, right, circuits. Each circuit is a flip-flop. The vacuum tube and transistor circuits are not as small as it is possible to build them, neither is the cryotron circuit. The size and complexity reduction is evident. (Martin Co.)

that a "wrap around" coil was unnecessary if the lead control conductor was made narrow and put very close to the tin gate conductor. The straight lead conductor acts like a coil consisting of half a single turn of wire. The current which flows along the lead conductor produces a magnetic field strong enough to destroy superconductivity in the tin gate and to operate the cryotron. This design simplified the cryotron so it could be much smaller and made to operate much faster. Success of the thin film cryotron now required development of techniques for making stencils or masks in order to provide the needed small and accurate lines. The close spacing of the lead control conductor over the tin gate conductor required a very thin insulation. This also had to be vacuum deposited and since it could contain no holes, it was difficult to produce. Progress is being made in these areas and gradually experimental cryotron circuits are being built that are smaller and that perform more satisfactorily.

The flip-flop circuit was used as an illustration of the cryotron in an electronic circuit. It was a simple circuit and has been widely used for remembering or storing information. While the cryogenic flip-flop still has many uses, a much simpler method of storing information was found. This is called the persistent current memory. Several different devices use this effect, including the Persistor, Persistatron, Crowe Cell, and Trapped Flux Memory. Each differs somewhat in construction and operation, but all have one thing in common. They store or remember information by means of a current which flows around a ring. The idea is the same as that of Collins' Experiment, when, you remember, he caused a current to flow around a lead ring for $2\frac{1}{2}$ years. Scientists realized that if a current can persist in a superconducting ring indefinitely, it can be used to remember information. The

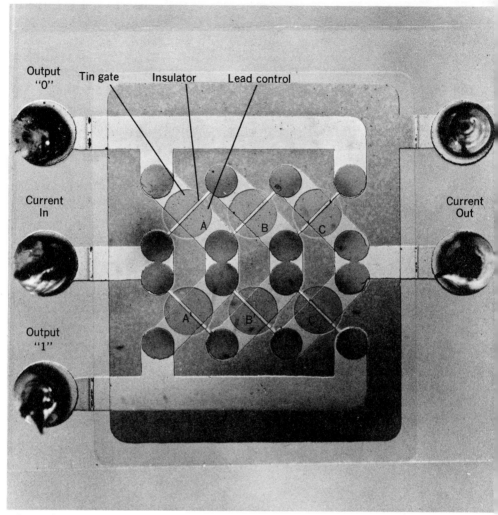

Output "0"   Tin gate   Insulator   Lead control

Current In   A   B   C   Current Out

A'   B'   C'

Output "1"

Three thin film cryotron flip-flops. They are connected one after another so that the last one is connected back to the first to produce an oscillator. (A. D. Little Co.)

direction in which the current flows can be used to distinguish between a "one" and a "zero." The presence or absence of the current may also be used. These persistent current

Superconductive memory circuit. A plate containing a number of memory devices is shown being lowered into a cryostat. (General Electric Corp.)

rings may be made just as small as the stencils will permit. Together with the conductors needed to start the current flow, and to measure it later, they can be grouped in large numbers to provide the memory capabilities needed for large computers.

The use of superconductivity has been applied to a number of electronic devices other than bolometers, cryotrons, and persistent current memory cells. In many applications, it is not necessary to be able to destroy superconductivity as it is done in these devices, but only to use the superconductive property itself.

One device of this type is the superconductive magnet. Although Onnes had not been able to obtain high magnetic fields, it was known that the electromagnet could produce weak fields. Attempts to increase the magnetic field by using purer metals, and making certain the crystal structure of the conductor was as good as possible, did not give much improvement. However, a few years ago, it was discovered that by using wire which had been deliberately mistreated and which had badly distorted crystals, superconductivity was obtained with very high magnetic fields. Superconducting magnets have been built which will produce magnetic fields of 70,000 gauss or more, which is about as large a field as can be produced by any large laboratory magnet. At first the niobium-tin alloy was used, since it has the highest known superconducting temperature, but while this worked well as a magnet, it was difficult to fabricate. Alloys such as niobium-zirconium and molybdenum-rhenium were developed which became superconducting in the 14–16°K range but which were easier to make and to wind on spools.

The superconductive magnets also use the effect of persistent currents. Once the current starts to flow and the desired magnetic field value is produced, the magnet can be switched by the use of a large cryotron to form a superconductive loop. The current and magnetic field will then persist with no additional supply of power to the electromagnet.

We get some idea how well the superconducting magnet compares with ordinary magnets if we look at the Westinghouse Corporation's description of its superconducting magnets. Westinghouse finds the superconducting magnet requires virtually no power to operate and less than 300 watts for a short time to start the field. The combined system of magnet and cryostat weighs only 200 pounds. On the other hand, the non-superconducting magnet which produces the same magnetic field requires 50,000 watts of power as long as it is operating, and it weighs over six tons.

So far the superconducting magnet only works with a direct current and produces a steady magnetic field. Much research is being directed toward discovery of a material which will remain superconductive while carrying a high alternating current. If one is found, it will be possible to build highly efficient transformers.

Another application of the improvement in performance given by superconductors is the superconductive delay line. For this we go from the direct current of the superconductive magnet up to the very high frequencies of billions of cycles per second used in radar and to the very high speed electronic computers.

It would be useful to be able to store the electronic signals at these high frequencies, sometimes for only a millionth of a second or less, to allow time for the radar or computer to perform some function. One way to provide this delay is to send the signal down a wire and wait for it to come out the other end. Since the signal travels with a velocity close to that of light, a long length of cable is required. It actually takes about 700 feet of wire to delay a signal one millionth of a second. Because there is an ever greater signal loss as the frequency increases, the line size must be increased to offset it. So it soon becomes impractical to build lines of sufficient length and size.

But if we use superconductive lines, the signal loss due to resistance in the conductors is eliminated. The loss in the insulation also becomes very small at low temperature, and so the line can be of very small diameter and still have very small losses.

The superconducting sphere, shown floating in space in the previous chapter, illustrates the reason for interest in a superconducting gyroscope. Gyroscopes use a rotating mass to indicate a reference direction. As long as no force is applied to this rotating mass, the axis on which it turns will always point to the same spot in space. However, even the best bearings for reducing friction in the supports of a conventional gyro are not perfect. They exert a force on the mass, and its axis of rotation drifts away from the reference. In a superconducting gyro, however, the mass can float in space, within a vacuum. With no supports to cause friction and exert force, the gyro accuracy should be greatly improved. Research being conducted on the superconducting gyro today should soon lead to devices for use in guiding ships and space vehicles.

Besides the devices we have discussed, and others benefiting from the phenomena of superconductivity, there are a number of electronic devices which make use of low temperature to improve performance. In many cases, this is possible because of the decrease in the thermal motion of atoms in the material being used. We defined temperature as being related to the amount of motion of the atoms. This motion also causes

Top: Superconducting magnet used with a cryostat. Bottom: Superconducting magnet that will produce magnetic fields of over 50,000 gauss. (Westinghouse Electric Corp.)

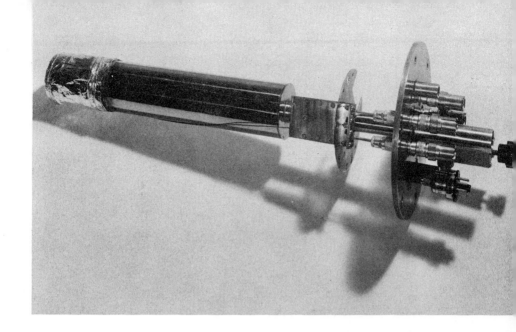

TOP: Superconducting delay line. BOTTOM: Two coaxial cable delay lines. (Martin Co.)

"noise" in the radio signals or disturbs the orderly arrangement of crystals needed for very sensitive electronic devices.

Both Masers and Lasers, tremendous scientific developments of the last few years, illustrate these non-superconductive liquid helium temperature devices. Some types of masers and lasers can be built to operate at warmer temperatures, including even room temperature. However, highly important characteristics of the injection laser and others can be obtained only at cryogenic temperatures. The Traveling Wave Maser illustrates very well the uses of cryogenics. This maser requires a large magnetic field, which must not vary over a distance of often several inches, and a superconducting magnet has been used to provide the magnetic field for this helium temperature maser. One design makes even greater use of unique low temperature properties by placing superconducting metal plates around the area in which the field is to be kept uniform. Since a magnetic field cannot penetrate a superconductor, the field cannot spread out as it would normally. This gives a very uniform magnetic field over the required length, and yet does not need a large volume magnet.

This last example of several low temperature effects used together to produce a device of unique performance capabilities, shows the beginning of a cryogenic electronic system. While it is often impractical to provide a cryogenic environment to improve the performance of a single device, once the low temperature is available, more and more devices may be designed to take advantage of it. Cryogenic electronics will never replace room temperature systems, or even a major portion of them. But it should take its place among the store of environments and techniques available to designers of electronic devices and systems, who, as developments continue, must constantly be on the lookout for new methods of dealing with the new problems that arise.

Chapter 10

PHYSICAL RESEARCH

WE HAVE GAINED some knowledge of cryogenics, and of the many ways in which it and its effects are put to use today. Among the future possibilities of this field of science, the importance of research into the laws of nature ranks high, and may do much to enhance our knowledge of the world we live in.

Cryogenics plays an important role in many of the investigations into nuclear physics and is helping scientists in learning more about the particles which make up matter. In chemistry it is useful in the understanding of many reactions, as well as in providing methods of obtaining reactions which would not occur at ordinary temperatures. Cryogenics is highly important in the field of electronics, particularly in the study of electronic properties of solid state materials. It is also involved in metallurgical research—where knowledge is needed on the reasons for defects in crystal lattices, for example, and means by which material properties can be changed and metals made stronger. In these and many other areas, cryogenics plays a role, either directly or indirectly, in the attaining or applying of new knowledge.

One of the largest areas of research involving low temperature is that of nuclear physics. Science is still trying to understand more about the composition of matter, about the various forces that exist in the atom, about the particles present in cosmic radiation, and in the nuclei of atoms. Temperature, which we have referred to several times as motion of the atoms, obviously can work to our disadvantage if we are trying to measure properties of these moving atoms and the smaller particles. We know that the use of low temperature not only reduces the motion of an atom about its position in a lattice, but also slows down other effects, such as changes in crystal boundaries, the ability of atoms to move or diffuse in solids, and the rates at which many other physical and chemical changes occur.

The liquid hydrogen "bubble chamber" has been used by nuclear physicists in the discovery of several atomic particles and should continue to play an important role in future basic nuclear research. The bubble chamber is used to provide a way of seeing the paths or tracks of nuclear particles coming from a particle accelerator. It also shows the tracks produced by the atomic fragments when one of these particles collides with one of the hydrogen atom's nuclei or its electron. By measuring the direction of the tracks and the amount by which they bend when a magnetic field is applied, scientists can learn much about the charge, mass, and velocity of the particles which produced the tracks. In this way the bubble chamber serves much the same purpose as the Wilson Cloud Chamber, but has a number of advantages.

These atomic particle tracks are made visible in the hydrogen bubble chamber in the following way. The particles from the accelerator, either protons or electrons, enter one end of the liquid hydrogen chamber. The chamber is connected to large pumps by a large pipe and a valve which can operate very rapidly. This allows the pressure of the gas over the liquid hydrogen to reduce quickly when the valve is opened. This, as we know, will lower the hydrogen boiling temperature.

In preparation for taking a picture of particle tracks, the pressure of the gas with the liquid hydrogen is increased until the liquid stops boiling. The liquid hydrogen becomes very quiet and there are no bubbles. The valve is then opened and the pressure decreases very rapidly. The liquid hydrogen will boil violently at this lower pressure, but it takes a split

Liquid hydrogen bubble chamber. This huge instrument is housed in a separate building adjacent to the Bevatron Accelerator. The Liquid Hydrogen Chamber is visible in the left center of the lower level. (University of California Lawrence Radiation Laboratory)

second for the bubbles to form and for the boiling to start. During the split second, the nuclear particles entering the chamber at high velocity collide with, or pass near, the hydrogen atoms which are about to go into the gaseous state and produce bubbles. As the nuclear particles travel through the liquid hydrogen they lose energy by these collisions or near collisions. This energy which is lost has the form of heat and causes the hydrogen atoms left in the wake of a particle to be at a slightly higher temperature than the undisturbed atoms. These higher-temperature atoms begin to form bubbles, and a line of bubbles marks the path of the particle. These tiny bubbles are made visible by lights at the top of the chamber. Just as these bubbles become large enough to see, but before they spread and cause violent boiling, they are photographed by stereo cameras. The chamber is then repressurized and prepared for the next operation. The entire time from opening the valve to taking the picture is less than a second.

Hydrogen is the simplest of all atoms. Its atomic make-up gives liquid hydrogen advantages over other bubble chamber liquids and over the cloud chamber. The hydrogen atom consists only of an electron and a proton, and so it is easier to determine what happens when the hydrogen is struck by a particle than with the more complicated liquids, such as the alcohol used in cloud chambers or the liquids of higher-temperature bubble chambers.

Tied in with the quest for more knowledge of the composition of matter is a need for deuterium, the heavy isotope of hydrogen. The deuterium atom, consisting of a proton, a neutron, and an electron, is found in very small amounts in ordinary hydrogen. Since isotopes are very similar chemically, it is difficult to separate the hydrogen and deuterium. However, since they do boil at slightly different temperatures

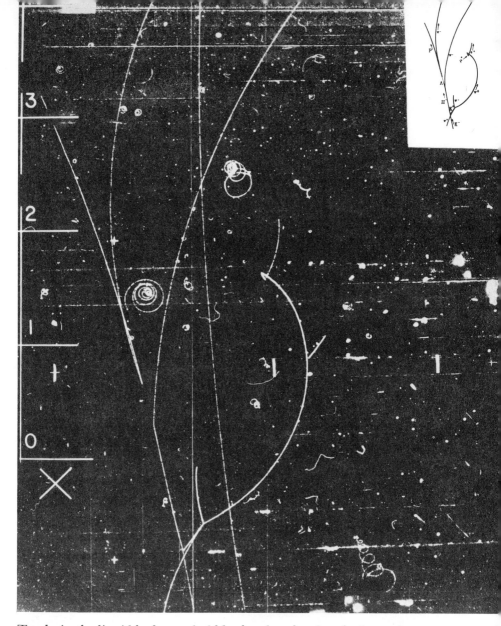

Tracks in the liquid hydrogen bubble chamber showing the formation, motion, and decay of sub-atomic particles. The small spirals are produced by electrons. The insert at the upper right shows the notation used in describing the particles producing the various tracks. (University of California Lawrence Radiation Laboratory)

(20.3°K for hydrogen and 23.6°K for deuterium), they can be separated by cryogenic distillation.

Perhaps the most significant research in cryogenic chemistry today concerns free radicals. A free radical is defined as any atom or molecule which possesses an unpaired electron. This does not include the ordinary stable molecules, such as oxygen, but does include hydrogen, chlorine, and the OH, CN, CH_3 molecules.

The free radicals are formed in many chemical reactions by the rupture of the bond in a stable molecule. They normally exist for only a very short time, during a reaction or in a flame, before they recombine. While they ordinarily last for less than a thousandth of a second, their properties are very important in learning the steps by which a reaction actually takes place.

At cryogenic temperatures the lifetimes of these frozen free radicals is much longer than at room temperature. This permits scientists to study them and learn more of the steps in complex chemical processes. They can even be collected on very low temperature surfaces and saved for study (Figure V-6).

While cryogenics is normally a region of very low chemical activity, some reactions have been found to occur only at these low temperatures while others do improve in speed or quality. The ability of charcoal, for example, to absorb gas improves considerably at cryogenic temperature. On the nuclear ship *Savannah,* a single pound of activated charcoal at liquid nitrogen temperature was found to remove as much radioactive krypton and xenon from the hydrogen gas surrounding the reactor as a ton of charcoal will remove at room temperature.

Research is being conducted into the nature of solids at low temperature. Many areas are being studied to learn more of

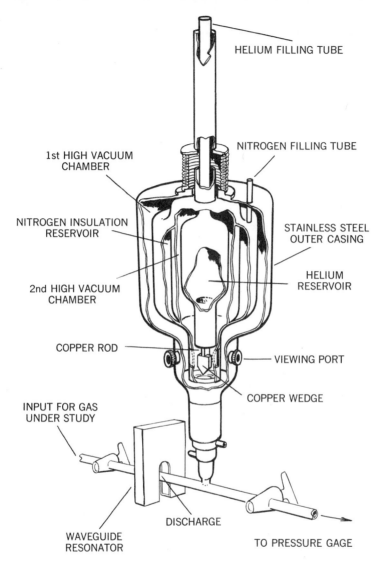

HELIUM FILLING TUBE

1st HIGH VACUUM CHAMBER

NITROGEN FILLING TUBE

NITROGEN INSULATION RESERVOIR

STAINLESS STEEL OUTER CASING

HELIUM RESERVOIR

2nd HIGH VACUUM CHAMBER

COPPER ROD

VIEWING PORT

COPPER WEDGE

INPUT FOR GAS UNDER STUDY

DISCHARGE

WAVEGUIDE RESONATOR

TO PRESSURE GAGE

Figure X-1. A cross-section of a typical stainless steel liquid helium research dewar built for free radical research. Radicals of the chemical under study collect on the copper wedge after being broken up by the waveguide resonator. Their characteristics during their short lifetime may be observed through the viewing parts. Note how the copper wedge is connected to the helium reservoir. The liquid nitrogen in the outer reservoir acts as a refrigerant for the liquid helium. (Sulfrian Cryogenics, Inc.)

the forces involved in solids, the make-up of crystals, and the many physical effects which occur in solids. Topics being investigated include such things as the change in the velocity of sound in a solid when it is compressed at low temperature, or the ability of molecules to vibrate together and resonate when receiving energy of just the right frequency.

This research has extended into the field of metallurgy. While the scientists are trying to learn more about properties of metals at low temperature, the engineers are already putting their discoveries to use. They have found that much greater strength is achieved if the metal is formed into its final shape while immersed in liquid nitrogen. Stress builds up as the temperature rises, which can be used to increase the strength of the item. This low temperature forming process is being used in making high-pressure containers with thinner walls than was possible before, for use in space craft.

One very powerful tool used in metallurgical research is the electron microscope. This microscope is similar in principle to its conventional counterpart, the ordinary optical microscope with which we are familiar. It is different in that it uses a beam of electrons rather than a beam of light to illuminate the object under study. By using electrons, a much higher power microscope can be built than is possible with ordinary visible light. This allows metallurgists to examine the crystal structure of their samples in very fine detail. The clarity of the picture which can be obtained with an electron microscope depends on the degree to which the electron beam can be focused. Focusing is done with a magnetic coil, which produces a very strong magnetic field. Electron microscopes are being designed which will use superconductive magnetic focusing coils. The large fields which these magnets provide will increase the magnification power of these instruments.

The use of the electron microscope is of course not limited to the field of metallurgy, but includes many areas of study, including biology and chemistry. All of these will also benefit from the superconducting magnet in the microscope.

The superconducting magnet at present finds its main use in research. Here it may be used to form part of new electronic devices, to study materials under high magnetic fields, or to assist in research with very high energy plasmas, as examples. Another example is its use in research with space propulsion systems. Here the magnet is housed in a special cryostat which allows hot gases to pass through the center of the magnet coil. While the magnet is aiding other research, research is also being done on the superconductors which permit the magnet to provide these high fields.

Our now familiar effect of superconductivity is under intensive study not only for application to electronic devices as we considered earlier, but also for the purpose of learning more about the effect itself, and thus about the properties of metals. Only recently Bernard Matthias at Bell Laboratories found that a mere trace of iron in some metals keeps them from becoming superconductive. This discovery reopens the possibility that all metals may become superconductive if the temperature is low enough and the metal is sufficiently pure.

The use of low temperature decreases the "noise" or static in radio receivers, permitting them to receive much weaker signals. While this is not an effective way of improving radios or television in our homes, it is very useful in special radio receivers which must be very sensitive. Very sensitive receivers are used in radio astronomy, for example. Here scientists study the weak radio signals received from the stars to learn more about the universe. While their main work is receiving and studying signals produced by nature, they are also listening for signals which may indicate the existence of other civilizations on planets beyond our solar system.

Many of the applications of the cryogenic environment depend on the ability to produce a low temperature conveniently and reliably where needed. This has led to the development of refrigeration systems which are not only small in size but also often "closed-cycle" machines. The term closed-cycle means that the gas, or liquid refrigerant, circulating through the refrigerating system is not allowed to escape but is reused. In this respect our household refrigerators are all closed-cycle machines. Like them, the closed-cycle cryogenic refrigerators

Closed cycle liquid nitrogen refrigerator. This one was developed to cool a detector for infrared (heat) radiation. The window which allows for the infrared radiation to enter is at the right. (Malakar Laboratories)

are used to cool objects to low temperatures and not as a source of liquefied gases.

Some closed-cycle liquid nitrogen refrigerators have been built which are small enough to carry in missiles and aircraft. The most common use of these refrigerators is to cool infrared detectors. The infrared detector measures the heat radiated from warmer objects, such as the exhaust from a jet engine, or even the heat from the surface of a missile. They are used to locate aircraft and missiles and to guide missiles to their targets. Low temperature operation permits use of more sensitive materials for the detector. Since it must be at a lower temperature than that of the object from which it receives heat, the detector is able to locate cooler heat sources. Closed-cycle refrigerators have also been built to keep electronic devices at liquid helium temperature. These are being developed by many different companies and are based on many different designs. One device which promises to reduce the size and cost of these closed-cycle helium refrigerators is the Turbine Expansion Engine, which is a form of the heat engine. It was used in nitrogen refrigeration many years ago by Peter Kapitza, a Russian physicist, but did not come into wide usage for helium refrigeration because it proved difficult to make. The lubrication of the rapidly rotating turbine shaft at low temperatures presented a problem. A method of separating the shaft from its support by a thin layer of gas looks promising. With this gas-bearing turbine, heat engines may soon be built which rotate at very high speeds and are about the size of the new high-speed dental drills. The development of reliable, low-cost, closed-cycle helium refrigerators will remove the main objection to cryogenic electronics ever since the time of Kammerlingh Onnes—the providing of the low temperature environment.

157

These various areas are but a few of those in which work is being done at low temperature. Many areas have been overlooked intentionally because the phenomena they are investigating are too complex to be described briefly. More have been omitted accidentally, since there are so many scientists working in the different areas of research involving low temperature that it is impossible to keep abreast of all of it. The days of the physicist who struggled to produce his cryogenic fluids before he could conduct his experiments are past. Now several hundred commercial helium and hydrogen liquefiers are in operation, as well as companies which produce cryogenic fluids and ship them to the users' laboratories. Cryogenics is no longer a curiosity. Its principles and techniques are being put to use in the research laboratories of almost all universities and large corporations.

We have seen many of the ways in which cryogenics has been applied. The vast amounts of research suggest that there will soon be many more. In our daily lives we have had very little direct contact with cryogenics in the past; we are unlikely to see it in use in the future. Our homes will most likely never have need for temperatures this low. But we are apt to see more and more large cryogenic tanks on the grounds of industries in the future and encounter more and more truck-cryostats on the highways. We may never get to see the liquefied gases, but we will see more of their effects. We have already watched television signals from Europe relayed from a satellite placed in orbit by a cryogenic rocket, weak signals which were amplified when they returned to the earth by a radio receiver at liquid helium temperature. How many other ways cryogenics will affect us indirectly can only be told by the future.

Index

Adiabatic, 69
Adiabatic demagnetization of paramagnetic salts, 68
ADL-Collins Liquefier, 126
Agriculture, cryogenics in, 47
Andrews, Dr. Donald, 127
Aneroid thermometer, 74
Argon, liquid, 33
Astronomy, radio, 155

Bardeen, Cooper, Schrieffer (BCS Theory), 122
Bimetallic strip, 74
Bistable Multivibrator Circuit, 131
Blood, preservation of, 42-44
Bogoljubov, N., 123
Bolometer, 127, 128
Bose, Jagadis, 109
Boyle, Robert, 20
Buck, Dudley, 128, 129

Carnot, Sadi, 71
Casimir, C. J., 122
Catalyst, 101, 102
Celsius, Andreas, 15
Chamber, bubble, 148
Chamber, Wilson Cloud, 148
Charcoal, 152
Circuitry, superconducting, 136
Coherence length, 125
Cold trapping, 34
Collins, Prof. Samuel, 51, 115, 116, 120, 128, 138
Cooper, Irving, 46
Conduction, 79
Conductivity, electrical, 97
Convection, 81
Conversion, Fahrenheit-Centigrade, 16
Creep, 108, 109
Crowe cell, 138
Cryobiology, definition, 38
Cryogenics in agriculture, 47
Cryogenics, definition, 1
Cryogenic distillation, 152

Cryogenic electronics, 126
Cryogenic fluids, definition, 22
Cryogenic medicine and surgery, 45
Cryogenic metallurgy, 154
Cryogenic temperature range, 1
Cryopumping, 34
Cryostats, 22, 28-33, 79, 82, 86-89, 107, 109
Cryotron, 128, 129, 133, 135
Cubic structure
  Body-centered, 93, 94
  Face-centered, 93, 94

Dalton's Law, 55
Delay line, superconducting, 142
Detector, infrared, 157
Deuterium, 150
Dewar, James, 49, 83
Dewar flask, 84
Dirac, P. A. M., 109
Distillation, cryogenic, 132
Droplet freezing, 40

Einstein, Albert, 109
Electronics, cryogenic, 126
Engine, heat, 63
Engine, turbine expansion, 157
Exchanger, heat, 58
Expansion engine, turbine, 157

Fahrenheit, Gabriel Daniel, 15
Faraday, Michael, 49
Fermi, Enrico, 109
Floating sphere, 120
Flip-flop, 131, 132, 135
Fluids, cryogenic, 22

Galileo, 77
Gas, general laws, 20; inert, 100; liquid, 22; liquification, 32, 33
Gay-Lussac, Joseph Louis, 20
Gorrie, John, 49
Gorter, H. B. G., 122

Ground State, 104
Gyroscope, Superconducting, 143

Heat, 12; engine, 63; exchanger, 58
Helium fountain, 107, 108
Helium I, 105
Helium II, 105
Helium 3 Isotope, 104
Helium 4 Isotope, 104
Helium, liquid, 103-106
Helium source, 30
Huggens, Charles, 44
Humidity, 55
Hydrogen, liquid, 100, 101

Insulating materials & methods, 85
Ionization, 14
Isotopes, helium, 104-105

Joule, James P., 50

Kapitza, Peter, 157
Kelvin, Lord, 58
Kelvin Scale, 16
Kinetic energy, 11

Lasers, 8, 146
Length, coherence, 125
Linde, Jonas, 49
Liquefier, ADL-Collins, 126
London, Heinz and Fritz, 122
LOX, 24

Machine, closed cycle, 156
Magnet, superconducting, 141, 142, 155
Mammoth, Beresovka, 41
Masers, 8, 146
Matthias, Bernard, 155
Medicine, cryogenics in, 45
Meissner, Walther, 119, 120
Meissner effect, 120
Memory, persistent current, 138
Memory, trapped flux, 138
Metallurgy, cryogenic, 154
Microscope, electron, 154
Missile, Polaris, 26

Newton's Second Law, 11
Newton's Third Law, 26
Nitrogen, liquid, 99
Nuclear spin, 101

Onnes, H. Kammerlingh, 51, 111-117, 126
Orthohydrogen, 101
Oschenfeld, R., 119, 120
Oxygen, liquid, 24, 99

Parahydrogen, 101-102
Paramagnetic, 65, 99
Parkinson's disease, 45, 46
Persistatron, 138
Persistor, 138
Pippard, Alfred, 122
Phonon, 124
Phonon vibration, 124
Pressure, partial, 56

Quantum theory, 109

Radiation, 81
Radical, Free, 152
Rankine scale, 16
Ray, Jean, 74
Refrigerating devices and systems, 51
Refrigeration, history, 48-51
Resistivity, 97
Rocket, nuclear propelled, 27; Saturn, 27; V-2, 24

Salts, adiabatic demagnitization of para-
    magnetic, 68, 69
"Savannah," 152
Seebeck, Thomas, 76
Semiconductor, 98
Silsbee, Francis, 117
Sound, first, 111; second, 110
Sphere, floating, 120
Superconductivity, 112-125
Superfluidity, 106, 107
Surgery, cryogenics in, 45

Tankers, methane, 31
Teflon, 94
Temperatures, boiling of common sub-
    stances, 23; physical, 10; relative, 10
Temperature range, cryogenic, 1
Temperature scale, Fahrenheit, 15; Cel-
    sius, 15; Kelvin, 16; Rankine, 16
Thermocouple, 75, 76
Thermometers, types of, 73-78; aneroid,
    74
Thomson, William, 58
Tudor, Francis, 49

Vaporization, latent heat of, 52-53
Vaporization point, 14
Van Braun, Wernher, 24

Wilson Cloud Chamber, 148

Zero, absolute, 9, 16, 19, 72; determina-
    tion of, 17-19

160